D0086743

Antoine Lavoisier—The Next Crucial Year

*

Antoine Lavoisier—The Next Crucial Year

OR

THE SOURCES OF HIS QUANTITATIVE METHOD IN CHEMISTRY

*

Frederic Lawrence Holmes

PRINCETON UNIVERSITY PRESS

PRINCETON, NEW JERSEY

Published by Princeton University Press, 41 William Street,
Princeton, New Jersey 08540
In the United Kingdom: Princeton University Press,
Chichester, West Sussex

Library of Congress Cataloging-in-Publication Data

Holmes, Frederic Lawrence.
Antoine Lavoisier, the next crucial year, or The sources
of his quantitative method in chemistry /
Frederic Lawrence Holmes.
p. cm.
Includes bibliographical references (p. -) and index.
ISBN 0-691-01687-9 (cloth : alk. paper)
1. Lavoisier, Antoine Laurent, 1743–1794.
2. Chemists—France—Biography. 3. Chemistry—France—
History—18th century. 4. Creative ability in science. I. Title.
QD22.L4H65 1997
540′.92—dc21
[B] 97-13566

This book has been composed in Galliard

Princeton University Press books are
printed on acid-free paper and meet the guidelines
for permanence and durability of the Committee
on Production Guidelines for Book Longevity
of the Council on Library Resources

http://pup.princeton.edu

Printed in the United States of America

1 3 5 7 9 10 8 6 4 2

* *Contents* *

✱ *Acknowledgments* ✱

THIS WORK was made possible because of the readily accessible archival materials of the *Fonds Lavoisier* in the *Archives de l'Académie des Sciences, Paris.* I am grateful to the Academy, and to its Conservateur, Christiane Demeulenaere-Douyère, for granting me authorization to include extensive excerpts from the first two volumes of Lavoisier's laboratory register, and from other archival documents. In accord with the policy of the Academy, I have reproduced in the endnotes the original French passages quoted in English translation in the text.

The two anonymous readers for Princeton University Press, and Bernadette Bensaude-Vincent, each made cogent suggestions for improving the original manuscript. Their comments prompted me to make several significant changes, including the addition of chapter 12.

Emily Wilkinson and Kevin Downing, both friends and colleagues as well as fine editors, made it a special pleasure to work with Princeton University Press, and the efficient copyediting of Bill Laznovsky helped move the manuscript smoothly into print.

In the Section of the History of Medicine at Yale University School of Medicine, my professional home for nearly two decades, my colleagues have given me necessary encouragement, and Patricia Johnson and Joanna Gorman have so perfectly managed the daily operations that surround me that I have been freed to concentrate the necessary attention to finish this volume in less than my accustomed time.

I began this manuscript in a motel in Raleigh, North Carolina, where I had accompanied my wife on a research trip for a biography of her grandfather. Since then, Harriet has encountered setbacks to her health which she has borne with a grace and courage that inspire all of us who share her life.

Frederic L. Holmes
December 1996

Antoine Lavoisier—The Next Crucial Year

*

* *Introduction* *

THE TITLE and subtitle of this book signify two themes interwoven into the texture of its narrative. I began to write about the first months in the investigative program that Antoine Lavoisier initiated in February 1773, because I thought that the distinctive quantitative chemical method with which he is historically identified could be seen emerging during those months. The evidence could be found, I believed, in the first of the twelve volumes of laboratory notebooks of Lavoisier preserved in the Archives of the Academy of Sciences in Paris. As I began to reconstruct the experimental activity recorded in that volume, my expectations were confirmed, but I saw that it provided also a compelling portrait of an energetic young scientist striving to overcome major difficulties at the beginning of the long investigative pathway to which he had committed himself. It seemed to me that I could incorporate into one narrative, both an account of the formative stage in the genesis of his "balance sheet" method, and the travails that Lavoisier underwent as he attempted to provide experimental support for a new theory that he believed, in his youthful enthusiasm, would shake the foundations of chemistry and physics.

To those familiar with previous Lavoisier scholarship it will be obvious that my title is intended also to suggest that the events described here form a sequel to those discussed in Henry Guerlac's important book *Lavoisier—The Crucial Year: The Background and Origin of His First Experiments on Combustion in 1772* (Ithaca: Cornell University Press, 1961).

I have previously written a longer book on Lavoisier's experimental investigations on topics connected with the chemistry of life. Much of it was drawn from the later volumes of his laboratory notebooks. Those experiments were often complicated, and my descriptions of them correspondingly intricate. The beginnings were simpler, even though, for Lavoisier, no easier. They were also more vivid and more poignant, because a younger, more impulsive, less experienced Lavoisier was seeking to exploit an accidental discovery that changed the course of his scientific life, but he did not yet know quite how to turn this singular event into a sustained investigative inquiry. During the early weeks of his venture, most of what he tried did not work. Yet he yearned to make public the findings he needed to support the new theory that he believed would initiate a new epoch in chemistry.

We shall follow Lavoisier day-by-day through these dramatic months. The descriptions of these first experiments through which his quantitative method took shape along with a nascent theoretical structure are not dull. His laboratory register records each experiment in the form of a little story. Nearly every one of these stories includes something that Lavoisier had not expected. If we follow his narratives with close attention, we can almost see this intrepid young man, talented and promising, but not yet proven, as he works in his laboratory setting. He is intense, impulsive, open to surprise, expressive of his aims, his hopes, his doubts, his disappointments, and his fears.

We shall also see a public Lavoisier announcing for the first time an "epochal" theory with a grand gesture and an aura of self-assurance that contrasts strikingly with the trepidations of the private investigator. The dissonance between the public presentation and the private state of the problems he faced may remind some readers of the differences between the public and private science of Louis Pasteur that Gerald Geison has so compellingly set forth. Yet the contrast between the older, powerfully positioned Pasteur, dominating his opponents, and a youthful Lavoisier seeking his place in the sun, is more striking than the similarities. It is easier for most of us to identify with the aspiring young Lavoisier than the mature, Olympian Pasteur. The experience of struggling to be ready for an event to which we have committed ourselves before we have realized how hard it will be to be fully prepared for that event, is one that this ambitious but appealing figure, not yet in 1773 one of the great heroes of the history of science, through the story told here, can now share with us.

. . . . .

The existing historical literature on Antoine Lavoisier is already very large, and has grown rapidly during the years between the bicentennial of his *Traité élémentaire de chimie* in 1989 and that of his death in 1994. The usual question, therefore—what room is there for still another book on Lavoisier?—must be addressed. The most direct answer is that Lavoisier is one of those relatively few, among the thousands of scientists of the past, who looms so large that he invites continuous historical reinterpretation. Lavoisier scholarship has been, and will always be, a collective enterprise. His career was so multifaceted, his impact so large, and so deep, that he will never belong to any single historian. Among the many valuable books and articles about him that have appeared recently, most are devoted to broader interpretations of his science, to the connections between his science and the other arenas of his life, or to the meaning of the revolution he led. My contribution, both in this volume, and my previous writing about Lavoisier, has dealt more with what I have called "fine-structure" studies of his science. If our understanding of the shape and significance of Lavoisier's life in science is to grow rather than merely to change, then there must be a continuing interplay between broad interpretation and attention to detail. There are still many documents in the Lavoisier archive that have been incompletely analyzed, stages in his investigative career that have been only partially interpreted. In my effort to probe more deeply into one year out of the quarter century in which he habituated the field and the laboratory, I hope to help sustain a stream of Lavoisier scholarship which has many miles still to flow.

Three years before the beginning of our story, Antoine Laurent Lavoisier had been, at the age of twenty-five, elected to the prestigious Académie Royale des Sciences de Paris as an adjunct chemist. He had received that early honor, not because of a singular scientific achievement, but through a combination of his assiduous application to diverse scientific activities, the impression of exceptional ability that he made on mentors such as Jean-Etienne Guettard, whom he had accompanied on geological-mineralogical surveys, the influence that his father and family friends could exert on his behalf, his own ambitious campaign to gain membership, and the perception that his inherited wealth would enable him to devote himself wholly to science. Once elected, he participated enthusiastically in the meetings of the Academy, served on its commissions, and prepared numerous reports, notwithstanding the fact that shortly afterward, his entry into La Ferme Générale, the private corporation holding a contract to collect taxes for the French Government, drew him away from Paris for much of the next three years. He also found time to perform experiments refuting the alleged transmutation of water, which drew considerable attention at the time and have been viewed by historians as a demonstration of the quantitative style of investigation that distinguished his subsequent chemical career from the qualitative practices of more traditional chemists.[1]

It is beyond the scope of this volume to summarize the events of Lavoisier's career prior to those that form our main subject. Readers encountering Lavoisier here for the first time can find them described accessibly in several recent biographies.[2] A succinct treatment of the events in chemistry preceding Lavoisier's entry into the field, and of the chemical revolution associated with his subsequent work, which manages effectively to integrate the results of more specialized scholarship up to the time of its publication, is the chapter on chemistry in Thomas L. Hankins, *Science and the Enlightenment* (Cambridge: Cambridge University Press, 1985). A thorough bibliography of recent scholarly writing about Lavoisier can be found in Patrice Bret, "Trois décennies d'études lavoisiennes," *Revue d'histoire des sciences* 48 (1995): 169–205.

In the following chapters I have described the chemical substances and phenomena investigated by Lavoisier and his predecessors in the language that they themselves used. To translate their terminology into the nomenclature that resulted from the chemical revolution would make reading easier, but would also distance us from the way Lavoisier and other chemists of the earlier period thought. Students of eighteenth-century chemistry are already familiar with this language. For readers who are not, I have included in an appendix a short discussion of the chemical terms used in the text, together with their more modern equivalents.

I have treated the first two volumes of Lavoisier's laboratory registers preserved in the Archives of the Académie des Sciences in Paris, on which much of the following narrative is based, as a full repository of his experimental activity during the period covered. There is no guarantee, of course, that they contain every experiment that he began or completed, but a reconstruction based on the assumption that they do does not lead to any conspicuous gaps or contradictions. Inserted in the notebooks are a few loose pages containing experimental data that is duplicated in more coherent form in the registers themselves. This evidence suggests that Lavoisier may customarily have made informal notes as he performed each experiment, then used these after the experiment was completed as a basis for the more complete record kept in the register. With a few minor exceptions, I have recounted all of the experiments Lavoisier performed after he began his extended research program in February 1773, until he completed a draft of the *Opuscules physiques et chymiques* in August of that year. For the period September 1773 until January 1774, I have described only those experiments relevant to the completion and publication of the *Opuscules.*

✻ CHAPTER ONE ✻

The Sources of Lavoisier's Quantitative Method in Chemistry

CHEMISTS, historians, and philosophers have long attributed to Antoine Lavoisier the introduction into chemistry of the quantitative methods characteristic of "modern" science. From an inconspicuous place in his *Traité élémentaire de chimie* they have extracted and made famous his statement that "nothing is created, either in the operations of art, or of nature, and one can state as a principle that in every operation there is an equal quantity of material before and after the operation."[1] This principle, treated by present chemical textbooks as "the cornerstone of all chemistry,"[2] is generally acknowledged to be far older than Lavoisier; but he is viewed as the historical figure who turned a general law of causality into the guiding principle of his research. "It is on this principle," he wrote, "that the whole art of making experiments is founded."[3] This statement had not been true when Lavoisier entered chemistry, but it was true of the mode of experimentation through which he himself transformed the practices of his field.

In the book which first popularized the phrase "Revolution Chimique" to summarize Lavoisier's achievement, Marcellin Berthelot wrote that Lavoisier's claim to be the principal author of that event rested on "three things, a fact, a theory, and a practice." The third, and most important of these—the practice—was "the exact weighing of all the products of chemical reactions: not only the weighing of solid or liquid products, as one had always done, but especially the weighing of gaseous products."[4] In the chapter on the conservation of matter in his *Identity and Reality*, the great philosopher Emile Meyerson wrote a few years later that "it was with the aid of the balance that Lavoisier accomplished his 'chemical revolution.'"[5] These judgments have been sustained by more recent historians. In 1961, Henry Guerlac affirmed in his nodal book *Lavoisier—The Crucial Year*, that "methodologically, the key to the Revolution was Lavoisier's systematic application of his 'special reagent,' the balance, not merely to solids and liquids, but also to the gases."[6]

Historians have recognized that it was not merely the practice of weighing gases, liquids, and solids that distinguished Lavoisier's chemistry, but its combination with the reasoning that Lavoisier summarized in the final sentence of his oft-quoted passage: "One must suppose in every case a true equality or equation between the principles of the bodies one examines and those which one obtains through an analysis."[7] Translating this idea into the modern language of chemical reactions, John McEvoy has written that "Lavoisier's quantitative calculations involved the mental use of equations in which the weight of materials remained constant during a reaction and only a change of arrangement took place." It was this underlying epistemological view that, according to McEvoy, distinguished Lavoisier from his contemporary Joseph Priestley, who also used quantitative methods and arguments, but who rejected the theoretical presuppositions on which Lavoisier based his practice.[8]

That the balance had many meanings for Lavoisier, both within and beyond chemistry, is the subject of a thoughtful chapter in Bernadette Bensaude-Vincent's important book on Lavoisier. I shall refer here only to two of her insights concerning its use in his chemistry. "Instead of concerning itself with the reactions themselves," Bensaude points out, the balance approach of Lavoisier "concentrates the attention on the initial and final states."[9] Lavoisier did not, in fact, use the term chemical "reaction." His favored term, "operation," did not distinguish between an intervention performed by the chemist and the interactions of chemical substances on one another. Bensaude notes also, that Lavoisier's balance method did not rely exclusively on weighings performed with the instrument known as the balance. She adduces the areometer, which Lavoisier used to measure the specific gravity of liquids, as an essential complement to his balance measurements. "The association of diverse gravimetric techniques," she writes, "created a new order of discourse," in which all substances were "rendered commensurable" through the measure of their weights.[10] To this central point I would add that Lavoisier's focus on gases made the most important complement to the balance his use of the pneumatic trough, invented by Stephen Hales, to measure the volumes of substances in the aeriform state. It was necessary, then, to know also the density of a given gas in order to convert a measured volume to a weight.

If there is widespread agreement on the importance of Lavoisier's quantitative methods, there is much less consensus about their source. Because pre-Lavoisian chemistry is assumed to have been primarily qualitative, some historians have looked outside of chemistry for Lavoisier's inspiration. In 1960 Charles Gillispie firmly rejected some of these alleged sources:

> Social scientists . . . sometimes like to see science drawing inspiration from society and politics. It has even been suggested that Lavoisier derived his chemical "philosophy of the balance sheet" from the accounting practices of the corporation which farmed the taxes. The economic interpretation of creativity takes special pleasure in Lavoisier's hobby of agricultural reform.

In his agricultural experiments Lavoisier recorded meticulously the quantities of seed, fertilizer, and work introduced into every field, and compared them with the quantities of grain harvested. "But surely," Gillispie objected, "the historian need not go thus far afield to find the origin of Lavoisier's input-output chemical procedure. Lavoisier . . . had studied Joseph Black before he became a gentleman farmer or a pillar of the tax farm."[11]

Gillispie's assertion that the model for Lavoisier was the quantitative reasoning applied by Joseph Black in 1755, in his retrospectively famous "Experiments upon Magnesia Alba," to prove the existence of "fixed air" relied on the assumption made by earlier historians of chemistry that Black's historic treatise, the opening signal for the events leading to the chemical revolution, was well known to all chemists by the time Lavoisier entered the field. This seemingly straightforward intellectual lineage has seemed more difficult to maintain since Henry Guerlac showed that chemists in Paris, including Lavoisier, were unfamiliar with Black's work at the time Lavoisier first took up his studies of combustion in 1772.[12]

Many historians have noticed that, several years before Lavoisier took up the questions about combustion and related processes to which he applied the "balance sheet" method, he had already used similar techniques and reasoning to show that water is not transmuted into earth.[13] This and several other youthful interests in instruments and measurements have lent support to a view that Lavoisier must have brought to chemistry a predilection for quantitative experimental methods acquired elsewhere. The assumption made by some earlier historians that the source for this approach was physics has been developed by Arthur Donovan into a vigorous argument that Lavoisier was inspired from the earliest stages of his scientific education by the precision and rigor of contemporary experimental physicists such as the Abbé Nollet, and that his driving motivation throughout his scientific career was to make chemistry more like physics.[14]

The need to identify such a specific source for Lavoisier's inspiration is diminished by the results of a recent set of collaborative studies which assert collectively that "from about 1760 on," there was a "rapid rise in the range and intensity of application of mathematical methods," including the adoption of precision instruments and methods of measurement, in many areas of activity, ranging from the physical sciences to economic and political management.[15] In accord with such an interpretation, a young man such as Lavoisier, whose formative years coincided exactly with this movement, would experience many inducements to prefer quantitative methods in any domain of activity to which they could profitably be applied.

Whether we choose with Donovan to affix Lavoisier's style of experimentation to his "attachment to the methods of experimental physics,"[16] or whether we allow greater latitude for the play of the "quantifying spirit" of his age on him, neither influence can explain the specific form of quantification embodied in the balance sheet method on which he built his revolution. I would like to suggest that, instead of seeking a source *from which* Lavoisier might have brought his distinctive method to the problem of the fixation and release of airs in 1772, we seek *within* his early efforts to cope with that problem the origins of what afterward grew into a method of far broader scope.

In *Lavoisier—The Crucial Year*, Henry Guerlac posed the critical question, how did Lavoisier "hit upon" the "idea of the role of air in combustion" that first led him to undertake his "classic researches"?[17] Since its publication in 1961, several other historians have offered refinements of the story Guerlac told, and the identification of a critical additional manuscript led Guerlac's student, Carl Perrin, to construct, during the 1980s a version which may well prove definitive.[18] The fascination with this question has partly arisen from the paucity of documentary evidence for the historical order in which Lavoisier made the initial moves that drew him into the problem. The first experiments which led him to make the claim, in November of that year, that phosphorus and sulfur gain weight when burned, and that metals gain weight on calcination, had also been sparsely described in the surviving documents from that period known until recently. The manuscript reassigned to that period by Perrin now gives us, however, full accounts of the two most critical of them. The evolution of a systematic method to attack these problems more comprehensively began after Lavoisier laid out for himself, in February 1773, a "plan" for a "long series of experiments" dealing with all of the processes he could think of in which "elastic fluids" are absorbed by or disengaged from solid and fluid bodies.[19] Fortunately, for this period the laboratory register which Lavoisier began to keep then provides much fuller documentation for his experimental activities than is available for the preceding period. The rest of this book presents the case that, by following this record of his investigation from his initial experiments on phosphorus and lead, through the first two volumes of his register, and by connecting what he did in his laboratory to the work of the predecessors on which he drew for both methods and ideas, we can witness the emergence, between November 1772 and October 1773, of the main outlines of what we have come to call his balance sheet method.

✻ CHAPTER TWO ✻

Consequences of the Crucial Year

WHEN Lavoisier described his first experiments on the combustion of phosphorus in a closed vessel, in a memoir he drafted on October 20, 1772, he was uncertain about how much weight the substance had gained. "This augmentation of weight, of which it is not easy to confirm exactly the quantity [proportion]," he wrote, derives from the combination of the air which is fixed in that operation."[1] Perrin has pointed out that the uncertainty arose from the rapidity with which the phosphoric acid absorbs water.[2] Lavoisier could not tell how much of the gain to ascribe to the water, and how much to the air. He could not even be sure that all of the increase was not due to the humidity of the atmosphere. What seemed for him for a time to be an "insurmountable obstacle" he overcame by an ingenious strategy. He burned 2 gros, 42 grains of phosphorus under a bell jar that he had saturated with water vapor. The product had the appearance of an acid very diluted with water. He placed it in a narrow tube and marked the height of the liquid. The liquid weighed 6 ounces, 7 gros, 69 1/2 grains. He filled the narrow tube to the same height with pure water, which weighed 6 ounces, 4 gros, 42 grains. The difference between them—3 gros, 27 1/2 grains— he reasoned, "could derive from nothing else than the acid contained in the water and lodged between its parts." The phosphorus burned (2 gros, 6 grains of the 2 gros, 42 grains with which he began) had gained more than one-third of its original weight. The result was a minimum, he realized, because if the phosphoric acid had actually increased the volume of the liquid, then his weighing of an equal volume of pure water had overestimated the weight of the water in the first liquid.[3]

This experiment can be seen in retrospect as the first one for which Lavoisier devised a complicated balance sheet approach to establish a chemical "operation." Until then, his effort to measure the weight gained by burning phosphorus was methodologically little different from that employed by Guyton de Morveau in the measurements of the weights gained in the calcination of metals that de Morveau had

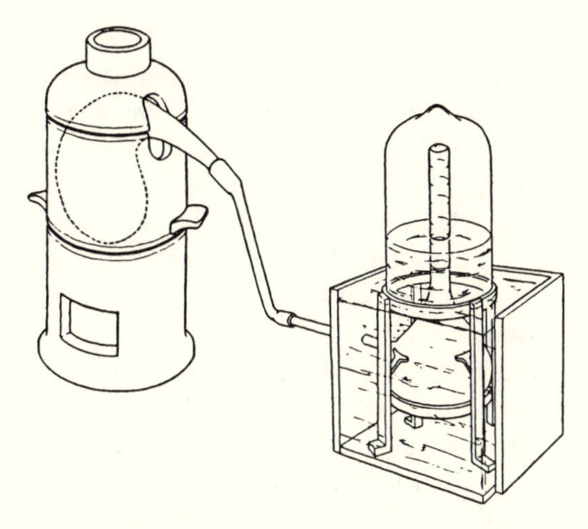

1. Guillaume-François Rouelle's modification of apparatus of Stephen Hales. Reproduced from Guerlac, *The Crucial Year*, p. 31.

announced six months earlier at the Academy of Sciences.[4] Lavoisier's innovation, as far as can be judged from his description, was not an application of a general strategy, but a response to a particular "obstacle" that he had encountered.

Believing that the weight gain "observed in the calcination of metals was too similar" to that in the combustion of phosphorus and sulfur not to "presume that they arose from the same cause," Lavoisier decided that "no experiment could be more decisive" for examining that question than "the reduction of lead calx in the apparatus of Hales." For this purpose he used a modified form of Hales's apparatus devised by Guillaume-François Rouelle, which consisted of a glass retort whose neck was bent in such a way that it connected with the interior of a vessel inverted over water. The air produced or absorbed in the retort would increase or decrease the volume of air in the vessel, lowering or raising the level of the water over which the air was contained. When Lavoisier heated a mixture of the lead ore minium with charcoal until the retort began to redden, "a prodigious quantity of air" was released, causing the water to descend rapidly. It rose again after the retort had cooled, but there was a net release of "462 cubic inches of air, that is to say, about one gros, which is nearly what lead gains by calcination." Noting that the material calcined had "lost only 2 gros of its weight," Lavoisier ascribed the difference between that amount and the one gros of air to "a little oil and phlegm which the lead had furnished during the operation and which had passed into the receiver."[5] He seemed unconcerned to verify that assumption, or to measure exactly the various quantities involved. In the "sealed note of November 1, 1772" on which historians have previously relied for a briefer description of the same experiment, Lavoisier did not report weights at all. He stated merely that he had "observed the disengagement of . . . a considerable quantity of air, and that it formed a volume at least one thousand times greater than the quantity of litharge employed."[6]

Asking why Lavoisier chose not to examine the weight gain in calcination by calcining lead, an experiment that would have been most analogous to the experiments he had just done on phosphorus and sulfur, Perrin suggests that it was merely because the reduction was more easily carried out, and "just as compelling logically."[7] An additional reason might be that Lavoisier knew it would succeed, because Stephen Hales had already distilled minium in his apparatus and obtained "34 cubic inches of air" from "1,922 grains, which was a cubic inch, of red lead."[8] Lavoisier had studied Hales's work carefully, and, as he himself acknowledged, it was through "repeating his experiments myself that I recognized that I could not doubt that the augmentation of weight [in combustions] derived from a portion of air absorbed."[9] The critical experiment on the reduction of minium which launched Lavoisier into his most important research program can itself be viewed as essentially a repetition, with modifications, of one of Hales's many experiments.

If we can discern retrospectively in the formative experiments that Lavoisier performed late in 1772, an incipient "balance sheet method," we can also see the extent to which his methods were improvised responses to the problems of the moment rather than systematic applications of a new mode of experimentation brought from somewhere else to them. In the first case, Lavoisier made accurate measurements and close calculations, because they were necessary to resolve indirectly a dilemma that had prevented him from giving an unambiguous answer to the central question he asked. In the second case rough numbers were adequate, because all that he sought to show was that large amounts of air were released in the reduction of the calx. The alternate way in which he presented the result in the sealed note suggests that he had not even recognized weights as more fundamental than volumes in interpreting the relation between the substances existing before and after a chemical operation. Moreover, the experiments themselves were closely modeled on those of others who had studied similar phenomena.

Lavoisier filled the first four pages of his laboratory notebook on February 20, 1773, with "reflections" intended to help him form his research plan. Historians have made famous the single passage in which he predicted that the importance of the subject would occasion "a revolution in physics and in chemistry." More significant than this somewhat fortuitously prescient statement, however, are the more concrete summaries of the immediate problems that Lavoisier defined for himself. "It is certain," he began,

> that an elastic fluid is disengaged from bodies in a great number of circumstances, but there are [divergent] systems concerning its nature. Some, such as [Stephen] Hales and his supporters thought that it is the air itself, the atmosphere, which combines with bodies, whether by the operation of vegetation and the animal economy or by the operations of art. They have presumed that this fluid cannot differ from that which we respire except by being charged with harmful or beneficial vapors, according to the nature of the body from which it is drawn. Some of the scientists who came after Hales noticed differences between the air disengaged from bodies and that which we breathe which are so great that they thought the former was a different substance, to which they have given the name fixed air. A third group of scientists thought that the elastic matter which escapes from bodies differed according to the substances from which it was drawn, and they concluded that it was only an emanation of the most subtle parts of the body, of which one can distinguish an infinity of species.

After remarking that the numerous experiments of Hales, Joseph Black, David MacBride, Joseph Priestley, and others were still not enough to form a "complete body of theory," Lavoisier added, "It is certain that fixed air presents phenomena very different from ordinary air. In fact, it kills animals which breathe it, whereas the latter is necessary for their survival. It combines easily with all bodies, whereas the air of the atmosphere combines, in the same circumstances, only with difficulty, or perhaps not at all."

Regarding everything that had been done on these questions before him as preliminary, Lavoisier proposed to "repeat everything with new precautions, in order to connect what we know about the air which becomes fixed in or disengaged from bodies with other knowledge about these bodies, [he then changed this last phrase to "acquired knowledge"] and form a theory." That task would, he predicted, require an "immense series of experiments."

Astutely Lavoisier sought to identify a particular omission in what his predecessors had done as a point of departure for himself:

> An important point that most of the authors have neglected is to pay attention to the origin of that air which is found in many bodies. They could have learned from M. Hales that one of the principal operations of the animal and plant economy consists of fixing the air by combining it with water, fire, and earth to form all of the combinations that we know about. They could also have seen that the elastic fluid which is disengaged by the combination of acids with alkalis or with any other substance nevertheless comes originally from the atmosphere, from which they might have been in a position to conclude that this substance is the air itself combined with some volatile part which emanates from a body, or that it is, at least, a substance extracted from the air of the atmosphere.

The "important point" outlined in this paragraph defined for Lavoisier the first stage in the research he envisioned for himself as he undertook his grand project:

> This manner of viewing my objective has caused me to perceive the necessity first to repeat and then to multiply the experiments which absorb air so that, knowing the origin of that substance I can follow its effects in all the different combinations.
>
> The operations by which one can arrive at fixing the air are vegetation, the respiration of animals, combustion under certain circumstances, calcination, and finally certain chemical combinations. It is by these experiments that I have believed that I must begin.[10]

We must be closely attentive to the details of Lavoisier's "February memorandum" to understand the relations between what he set out to do then, what he had done in the preceding "crucial year," and what eventually emerged from the research trail begun here. The famous phrase notwithstanding, this is not the blueprint which foreshadows the broad outlines of the chemical revolution. In some dimensions it is even broader, but in others far narrower. Nor is it a simple continuation of what he had begun in the fall when he discovered that phosphorus and sulfur gain weight when they are burned. Moreover, the problems outlined here bear no explicit connection to the earlier speculations about the nature of the gaseous state on which historians have recently placed such emphasis.

In the manuscript "On the cause of the weight acquired by metals and some other substances by calcination," which Lavoisier wrote, according to Perrin, in November 1772, he had acknowledged that he had been led to his dramatic experiments on the combustion of phosphorus and sulfur, and the reduction of lead ore, "more by accident than by theory." The conclusions he drew, that in combustion and calcination substances gain weight by absorbing "air," that "air enters materially into the composition of metallic calxes" and is released in their reduction, led Lavoisier immediately to consider that the theory of Stahl might require modification. That theory had "regarded every calcination as a loss of phlogiston, whereas it is [now] proven that there is simultaneously loss of phlogiston and absorption of air."[11] As Guerlac and Perrin have both stressed, Lavoisier's purview of the problem was, at that time, relatively narrow. He referred to older authors from Boyle to Hales and Stahl, and some more recent ones, but without awareness of the issues raised by Black's discovery of fixed air.[12] He referred to the "air" absorbed or released without questioning its identity. By the time he outlined his research plan in February, all that had changed. His reading of the publications of Black and subsequent authors who had treated the subject had led him to believe that the identity and origin of the air which combined with "bodies" was exactly the central question. For the moment, at least, the questions about combustion, calcination, and the role of phlogiston that had recently preoccupied him were thrust into the background, or subsumed within a larger problem. There remained room in his agenda for the study of combustion and calcination, but now merely as two examples of processes which absorb air. In his research plan they appear consistently less prominent than the "operations of nature"—vegetation, respiration, and fermentation—through which he imagined the air of the atmosphere was mainly combined with other bodies.

* CHAPTER THREE *

Vision and Reality

THE FIRST experiments Lavoisier recorded in his notebook—"experiments to make fixed alkali caustic"—he dated February 23, or two days after writing out his research plans. He had probably started them, however, at the beginning of February. They were based directly on Joseph Black's description of the action of lime on fixed alkali: "If quick-lime be mixed with a dissolved alkali, it likewise shews an attraction for fixed air superior to that of the alkali. It robs this salt of the air, and thereby becomes mild itself, while the alkali is consequently rendered more corrosive."[1] Lavoisier chose not to use the most common fixed alkali, salt of tarter, because its avidity for water made it difficult to keep at a fixed point of desiccation.[2] Instead, he mixed purified crystals of soda with quicklime in water. He did not obtain exactly what he had expected. After changing the proportions and filtering, he received an alkaline fluid which did not effervesce with acid; but after adding more water, filtering again, and allowing the filtrate to stand for fifteen days, he "saw with astonishment that [the alkaline liquid] made a lively effervescence" with acids. He could think of two explanations: either he had not used enough quicklime, so that he had obtained a mixture of caustic and mild alkali, or the alkali had, during thirteen days of exposure to the air, reabsorbed air. To decide between these two "opinions," he shook some of the alkali obtained from the second operation in a matras, but this time it did not effervesce with acids. He then heated the solution "to test whether the air would combine more easily, but the matras broke."[3]

In these experiments Lavoisier appeared mainly to be examining for himself the operations through which Black had discovered the "relations between fixed air and alkaline substances."[4] His early surprises and mishaps may have taught him that even the repetition of the experiments of those whose views on fixed air he wished to reassess would not be uneventful.

The first experiment that Lavoisier *began* after laying out his research project was the "calcination of lead in a retort," on February 22. It can be viewed either as the beginning of his newly planned series of experiments on the "operations by which one can fix air," or as the continuation of his experiments of the previous fall to determine the cause of the "augmentation of weight that one observes in the calcination of metals."[5] Lavoisier himself need not have noticed the distinction. The difficulties into which he quickly ran are prominent in his lively narrative of the event:

> I took 2 pounds of well-purified lead and cut it into very thin strips which I divided with a scissors. I introduced them into a clay retort and connected the Hales apparatus. I heated gently, then I pushed the fire strongly for four hours. The bars which supported the retort became absolutely red. The air expanded during the entire operation, but what gave me the most anxiety was that even after the fire had reached its highest degree the water kept always falling. I stopped [the fire] and the air soon began to contract and the water to rise. I was very surprised when, at the end of 2 hours I saw the water soon descend again instead of rising. The retort being checked, it had a small, imperceptible crack, through which the air had entered.

The lead had been calcined on its surface. Lavoisier wished to repeat the experiment, but he "fell again on a retort that admitted air." He had luted all the joints of his apparatus so carefully that the retort was the only possible source of the leak that again ruined his effort. A third experiment yielded only the same result as the first.[6]

"These repeated accidents obliged me," Lavoisier related, "to have resort to glass retorts." He decided at the same time to calcine tin instead of lead. He prepared the metal in the form of small granules by melting it and letting it fall into water. After placing twenty ounces into the retort, he heated it gradually until the air in the pneumatic vessel began to expand. He increased the fire "insensibly," but the water soon ceased to descend. Lavoisier tried to keep the fire constant, but he "augmented it in spite of myself, and I saw the water begin again to descend, which made me fear that the retort had collapsed, and that is, in fact what had happened. It was entirely deformed, and on cooling it cracked and air entered it, thus the experiment failed." Noticing that the tin was less calcined than the lead had been, Lavoisier decided that lead was more suitable for calcination in closed vessels.[7]

After this series of fiascos, Lavoisier tried a completely different approach to the study of airs fixed in bodies. Borrowing a pneumatic pump from the Academy of Sciences,[8] he attempted to draw air from alkalis by subjecting them to a vacuum. One of the alkalis gave off a few "imperceptible bubbles," but otherwise there was no effect except a strong odor of volatile alkali. Lavoisier concluded that it was not air that evaporated, but "another fluid which we do not see." In another experiment he placed volatile alkali and minium under the receiver of the pump. As soon as he began to evacuate it, a violent bubbling began, and the air which came out of the pump had such a penetrating odor that it nearly suffocated him. The volatile alkali seemed to him to have "all the characteristics of causticity," from which he inferred that "it seems, therefore, that it is deprived of air." That observation plunged Lavoisier into doubts about his nascent theoretical structure. "How," he asked himself, can that state of the alkali be reconciled with the presence of the minium?[9]

> The latter substance is, according to me, a combination of air and lead. That air ought to pass into the volatile alkali, with which it has a prodigious affinity. Nevertheless, it dissipated and escaped during the operation.
>
> This difficulty is puzzling, and I believe that one must conclude that the air combined with lead in minium is not at all the fixed air which has such an aptitude to combine with alkaline bodies. It is undoubtedly the air of the atmosphere. Perhaps also this air which is disengaged from lead is not sufficiently charged with phlogiston to combine with alkalis. Because according to the system that some hold, fixed air is air combined with phlogiston. But I admit that all of this presents great uncertainty.[10]

The first week of his new experimental campaign had led Lavoisier only into technical failures and conceptual conundrums. Reassessing his situation, he decided once again to change his approach. Instead of performing further experiments with the ordinary laboratory apparatus that had repeatedly let him down, he now began to design new types of apparatus to study the properties of fixed air.

On March 1st, Lavoisier wrote into his laboratory register descriptions of several pieces of apparatus that he hoped to have fabricated. The first was an "apparatus to combine fixed air with different species of liquid." It consisted of a siphon through which the air passed into the liquid contained in a jar, then exited through an orifice. The air could not escape from the jar without traversing the liquid, and one could continue passing it through until "the combination would be made." Lavoisier added to the design a funnel, fashioned from metal, so that the apparatus could be joined to a furnace to take up vapor produced from charcoal.[11]

The rest of the apparatus Lavoisier described indicates that he intended to direct his attention toward the physical and physiological properties of fixed air. "A primary consideration in the study of fixed air," he wrote, "is to determine its weight. Next it will be necessary to examine whether it is elastic as the atmosphere is. In the third place, to what degree it is compressible. Fourth, to wash it with different materials in order to observe afterward its effects on animals."[12]

For weighing the air Lavoisier imagined "a simple method," not previously known, which would "eliminate all errors on the part of the balance." The apparatus consisted of a large spherical glass vessel to which was affixed a copper valve through which it could be connected to a vacuum pump. After filling the vessel with the air and weighting it so as to be "equiponderable with water," one would evacuate it, close the valve, and replace the connecting tube with one to which weights could be added. The amount of additional weight necessary to make it again equiponderable with water would give the weight of the air. "The principal object here," Lavoisier added, was less to determine the absolute weight of the air than the "relative weights of different airs." He realized, however, that he could reach the same goal in a "simpler, less expensive way," by means of a vessel which would first be filled with ordinary air, and suspended in water. Lead weights would be added until the surface of the water came to a mark on the vessel. One would then displace the air by passing fixed air into the vessel through a syphon, and readjust the weights until the water came to the same mark. "The difference between the weights will give the relation between the weight of fixed air and ordinary air." The method was obviously an adaptation of the areometer methods Lavoisier had long used to measure the specific gravities of liquids. "The difficulty," he quickly realized, "will be to dry the vessel well."[13]

To measure the "elastic force" of fixed air, Lavoisier thought that a simple barometer would suffice, and that he could introduce the fixed air by means of a small syringe. It would be more difficult to measure the compressibility of the air. A normal "machine to compress air" could be used, but he devised a special bent tube with a narrow opening to displace the ordinary air with fixed air.[14]

"The machine to test the effects of air on animals," Lavoisier noted, had already been ordered, and was nearly completed. He therefore did not describe it in his notebook, but he reminded himself, "in the experiments on animals, do not omit frogs."[15]

None of these plans worked out. The workers Lavoisier had commissioned to build his apparatus did not come through. While waiting for them to finish, he became sick and could not continue his research. By the end of March, having lost whatever patience he may have had, he contrived a hasty expedient to make some progress before the general meeting of the Academy of Sciences at which he had hoped to present the first results of his new endeavor:

> The various machines described above have not been completed, due to the slowness of the workers, [and] an illness of fifteen days and other business has, moreover, forced me to interrupt my experiments. Nevertheless, as I would like to have something to announce at the public meeting [rentrée], and time is pressing, I imagined that one could do some of the same experiments in a simple way with the burning glass.[16]

The burning glass was the great Tschirnhausen lens that members of the Academy had borrowed from that institution's collection of curiosities during the preceding summer for experiments on the burning of diamonds. Lavoisier had been one of the several Academicians who carried out these experiments during the fall. They had used the cumbersome instrument then in a shed located in the Jardin de l'Infant of the Louvre.[17] Now, however, Lavoisier managed somehow to install the lens in the "Queen's apartment," inside the Louvre, where he could perform his experiments under more constant temperature conditions.[18] His plan was to attempt the calcination and reduction of metals inside a pneumatic vessel, heating the material by shining the rays of the sun through the glass walls of the vessel so that they would come to a focus on the metal or calx employed. In his anxiety to have something to report within the seven weeks left to him before the public meeting of the Academy, Lavoisier thus retreated from his broader program for the study of the absorption and release of airs to the core of his prior concern with the cause of the weight gain in calcination.

Too much in a hurry now to wait for specially designed apparatus, Lavoisier put his together from easily available objects. For the inverted vessels he procured a supply of small glass jars, ranging from five to seven inches in diameter. A rectangular china basin used ordinarily for washing hands made do nicely for the water trough in which he immersed the jars. At a china-ware dealer he picked up a crystal pedestal normally used to hold fruit, which served perfectly to support the metal or calx above the water inside the vessel.[19] The knowledge of fixed air that he had acquired since the beginning of the year by reading the earlier literature made him realize that he could not contain it in a vessel directly over the water, because "fixed air has the property of combining with water." He solved this problem simply by covering the surface of the water with a layer of oil which he could introduce into the vessel with a syphon. Because oil does not absorb fixed air, this maneuver enabled him to "conserve all of the quantity of fixed air that is disengaged."[20]

With his quickly assembled apparatus Lavoisier attempted, on March 29, to carry out the calcination of lead that his failing retorts had prevented him from achieving five weeks earlier. After placing 3 gros of very thin curled strips of lead into a hard clay dish supported on the pedestal, Lavoisier raised the water into the vessel with a syphon, marked its level on the jar, and "presented the apparatus to the burning lens." The lead soon began to melt, becoming covered with a layer of yellowish calx. As he continued the operation, however, he "saw with surprise that the lead did not calcine further." That unexpected obstacle prompted some thoughts about the nature of the process:

> I began then to suspect that the contact of circulating air would be necessary for the formation of that metallic calx: even, maybe, that the totality of the air that we respire does not enter into the metals that one calcines, but only a portion which is not present in abundance in a given mass of air. Perhaps also the layer of calx which covers the surface of the metal prevents direct contact with the air and stops the progress of the calcination.[21]

Of these three possible explanations, one represented a potential advance in Lavoisier's general understanding of the relation of the airs fixed in bodies, while the other two pointed only to technical problems. He was, however, not in a position to know which explanation was right.

Despite his "troublesome situation," Lavoisier kept the lead in the focus of the burning glass for a long time. The air, which had expanded during the heating, contracted as he allowed the vessel to cool. The next morning he found the water level two lines above the original

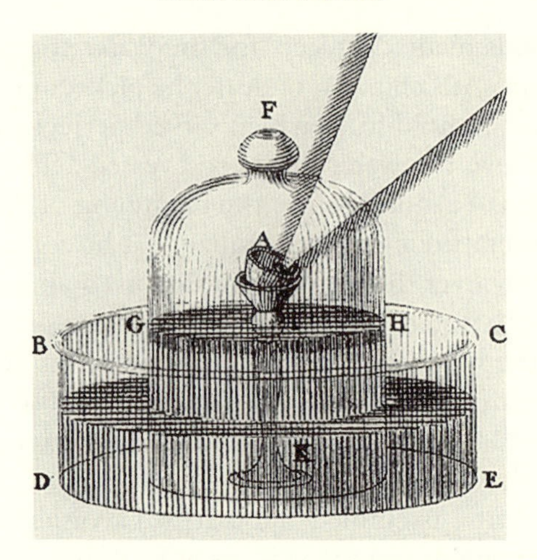

2. Apparatus assembled by Lavoisier for experiments with the burning glass. Reproduced from *Opuscules*, Plate II, Fig. 8.

mark. Although he had been aware of the need to determine the effects of changes of temperature on the expansion and contraction of airs, he was not concerned that such effects might interfere with this experiment, because the temperature in the room had not changed sensibly during its course. To enable him to convert changes of level in the vessel easily to volumes, he had measured the circumference of the jar and written out a table of the equivalent in cubic inches to each line of height. The two lines give, he wrote "almost 7 cubic inches" (6 2/3 according to his table), "that is, nearly one-quarter of a grain." (It is not clear how he estimated the conversion to weight, since he had not carried out his planned measurements of the weights of airs.) At this point the outcome appeared successful in spite of the limited calcination, but Lavoisier noted that he should "nevertheless, repeat the operation." His hopes for this one collapsed when he removed the lead and found that it had not gained weight. "On the contrary, it lost about half a grain, no doubt because of the vapors that it exhaled." The abundant fumes given off during the operation had formed yellowish flowers on the top of the jar, but he did not try to save the experiment by collecting them.[22]

Thinking he might overcome his troubles by trying another metal, Lavoisier put some zinc under another jar and again applied the burning lens. The calcination appeared to go well, but after the operation was over, the surface of the liquid returned to exactly the level at which it had begun, and when he retrieved and weighed the zinc he found it to have lost a quarter of a grain.[23]

Frustrated once more in his calcination attempts, Lavoisier turned back to the only operation with metals at which he had previously succeeded, the reduction of minium. Maybe the simplicity of his new procedure could yield more reliable results than had the experiment carried out in the fall with a retort and Hales apparatus. Placing 2 gros of minium and 1/2 gros of powdered charcoal in the vessel, he began heating, and noticed immediately small globules of lead appearing on the surface. Although he continued to expose the material to the focus of the lens for a long time, the reduction did not proceed beyond the surface of the calx. The water level fell by 5 1/4 lines, corresponding to about 9 cubic inches, or "almost 5 grains." That seemed to Lavoisier to be too much, in view of the small quantity of minium reduced, and he "feared that the charcoal had contributed some" of the air. The experiment thus turned out to be yet another disappointment.[24]

Undaunted, Lavoisier tried on March 30 to calcine tin, but again found the water level at the end of the operation just where it had begun, and the superficial calcination to have increased the weight of the zinc by only "a sixth or an eighth of a grain." Perhaps to check his concern that the charcoal in his last reduction experiment had supplied some of the air, he next placed a mixture of charcoal and well-calcined fixed alkali into another of his jars and heated it as usual. Nothing was absorbed or released; but, he remarked, "this experiment does not prove anything directly, because the alkali, . . . being very deprived of air, and consequently very avid, might have absorbed that which was formed" from the charcoal.[25]

In another effort to obviate the possible contribution of air from the charcoal, Lavoisier repeated the reduction experiment with the same amount of minium as before, but with one-third as much charcoal (12 grains in place of 1/2 gros). The sun was very bright that day, and the reduction went "promptly," yielding large globules of lead. A difficulty arose in the form of some yellow vapor that attached itself to the top of the jar. Lavoisier thought this was probably volatilized calx, and worried that it had "probably absorbed air." When all was over the next morning, the water level was only "3 1/2 lines, at most," below its original position. After determining the diameter of the jar at that level and deducting 1 cubic inch for the volume of the pedestal, Lavoisier estimated that the "quantity of air disengaged was 13 1/2 cubic inches."[26] The fact that he did not fill in the space left in his notebook to record the weight of the lead produced suggests that Lavoisier was probably not entirely satisfied with the experiment; but when compared with the mishaps that had attended each of the preceding experiments, this one he must have regarded as at least a modest success.

Having an idea that "the calcination of tin and lead might operate more easily" than that of either metal alone, Lavoisier composed an alloy by mixing together one gros of each metal. He seems to have had no better reason for his belief than the pure hope of finding some way to do better than he had up until then. Nevertheless, the event seemed to justify the effort. The alloy melted immediately, and "many white fumes arose, attaching themselves partly to the top of the jar, and depositing partly on the surface of the oil." After only a quarter of an hour he judged the calcination to be "almost complete." By evening, 3 1/2 lines, or 5.74 cubic inches of air had been absorbed, although the next morning the amount had diminished to 4.92 cubic inches. Despite this positive result, the large quantity of fumes made him uneasy. He feared that in this case, as in the previous calcinations, the intense heat at the focus of the lens had volatilized some of the metal, making it impossible to obtain a "fixed result for the augmentation" of its weight. Probably for this reason, he put the calx aside without bothering to weigh it. Another possible source of error in this and the preceding calcinations that worried him was that exposure of the surface of the oil to the heat might have caused it to produce or to absorb air.[27]

Lavoisier made one more effort to calcine zinc, but once again found it impossible to "carry the calcination further" than a surface pellicle. Observing the water level the next morning "with much attention," he still could not be certain whether any air had been absorbed, or whether a rise of about one-quarter of a line was due to a slight change in the temperature.[28]

Still concerned about the air that might be released by the charcoal used for the reduction experiments, Lavoisier now placed 36 grains of powdered charcoal alone in the usual manner under a smaller glass jar. He had only begun the operation when the sun became obscured by clouds. He had just time to see the charcoal redden and begin to decrepitate in a "remarkable spectacle." Despite the interrupted operation, the charcoal produced almost 2 cubic inches of air.[29] That result could only reinforce his apprehension that in the reduction of minium some of the air produced derived from the charcoal rather than from the lead calx.

While he was engaged in these experiments, Lavoisier looked up the specific gravities of lead, tin, and zinc, in "the tables of M. Cotes." From these figures he calculated the weight of a cubic foot of each metal. The 2 gros of lead that he had used in his experiments, he found, equaled 0.0336 cubic feet. The same weights of tin and of zinc were equal, respectively, to 0.0516 and 0.0542 cubic feet. No published tables gave the specific gravity of minium, but from a statement that litharge weighs "six times as much as water" he estimated that 2 gros of the lead ore would be about 0.127 cubic feet.[30] Why did Lavoisier want to convert his measured weights of these substances to volumes? As can be seen from the descriptions of some of these experiments that he later published,[31] he did so in order to report how many times the volume of air absorbed or released from the body was to the volume of the body that contained it. In doing so he was following the example of Stephen Hales, who regularly reported the relative "bulks" of the air and the bodies that generated it. Hales did so to dramatize the "vast contraction and expansion" that the air underwent when it was fixed in and released from bodies.[32] Did Lavoisier follow Hales blindly in this respect, or did he have his own motivation for wanting to present his results in that form? The most we can say is that his doing so indicates that, at this early stage in his investigation, he had not yet settled on the comparison of weights as the fundamental principle of his quantitative methods.

Ever since the experiment of February 26, in which he had placed minium and caustic volatile alkali together under a vacuum, Lavoisier had worried about the theoretical consequences of the failure of the air contained in the minium to combine with the alkali. Apparently unpersuaded by his own conjecture at that time that the air in the minium was different from fixed air, he saw himself faced with a crisis in his theoretical "system." On April 6, he put down in his laboratory notebook the reason for his concern:

> I have many times made very strong objections to myself against my system of metallic reductions, and here is what it consists of. According to my view quicklime is a calcareous earth deprived of air. Metallic earths are, on the other hand, metals saturated with air. Nevertheless, the one and the other produce a similar effect on alkalis, they both render them caustic.

Lavoisier had, in effect, accepted Joseph Black's theory that caustic alkalis are produced by removing fixed air from ordinary alkalis. To fit his own theory of calcination to that view required, Lavoisier thought, that minium should remove the causticity of an alkali by restoring fixed air to it. That minium seemed to have the opposite effect was the source of his anxiety. To situate his crisis more firmly, he decided to "verify the fact" that minium renders alkalis caustic. So far he had inferred that action mainly from the fact that a caustic volatile alkali had not absorbed the air from minium. First he boiled 12 ounces of crystallized soda with two pounds of minium, but found that the alkali had not become caustic. Not satisfied, he boiled a small portion of the same liquid with another half pound of lead, adding fresh water to avoid desiccation, and still found that, like ordinary soda, the alkali precipitated with lime water. Applying one more test, he divided a solution of mercury in nitrous acid into two portions. Into one he put several drops of a solution of soda, into the second he dropped soda previously passed over minium. Both produced precipitates, but the second was much darker than the first.[33]

After obtaining these results, Lavoisier recorded with relief that "the objection seems to me completely destroyed with regard to fixed alkali." He had, however, not fully resolved his crisis, because the difficulty "subsists in its entirety for volatile alkali":

> In fact, the volatile alkali that one obtains from a combination of sal ammoniac and minium is deliquescent, does not precipitate lime water, is infinitely penetrating. It is, therefore, evident that it is in a caustic state. How to explain this phenomenon? I admit that I still know nothing about it.[34]

Seeing no way out of this impasse, Lavoisier took up on April 9 a different combustion phenomenon, the detonation of niter and sulfur. He placed a mixture of 4 gros of sulfur and 2 gros of niter under one of his jars and tried to detonate it with a magnifying glass. His string of misfortunes seemed to continue, however, because no detonation took place, only a brief combustion, and after a short time "all was extinguished." He kept trying, but was not able to burn even the sulfur. Rather, it was reduced to flowers which sublimed. The operation had absorbed 19 cubic inches of air, but after several days he found that the water had returned to its original level. He exposed the mixture to the large burning lens, but still could obtain no more than the beginning of a combustion, with no detonation. He let the jar and its contents stand for several weeks.[35]

While Lavoisier coped with one setback after another, time was running out for him to make a significant advance that he could report at the public meeting of the Academy. Theoretically, his "system" was in disarray, and six weeks of experimental effort had produced little but failure. The most he had to show for it was a partially successful reduction of minium with the burning glass, which only confirmed the result he had already reached by another means in November.

✻ CHAPTER FOUR ✻

The Public Arena

IT IS a measure of both the audacity and the ambition of the young Academician, not yet quite thirty years old, that in the midst of this sea of troubles Lavoisier not only carried through his intention to present "something" to the public at the open Easter meeting of the Academy, but made it the occasion to announce in sweeping terms his new "theory." In the paper he prepared, he aimed to embrace, even in the "brief time" allowed him in this prominent forum, both his own initial discoveries about combustion and calcination, and some of the broader range of topics that he had set out to examine in February.

Historians have regarded Lavoisier's "Easter memoir" as his *first* public announcement of his theory. The note he had deposited on November 1, 1772, at the Academy remained sealed,[1] and, according to Perrin, the memoir that he had prepared for the public meeting held in that same month remained unread.[2] The question remains open, had he revealed nothing to his colleagues about his new ideas during the intervening four and a half months, or did he only present formally on April 21 what he had already shared with those around him?

In his sealed note Lavoisier had written that his reason for depositing it was that the discovery he had made appeared to him the "most interesting that had been made since Stahl, and . . . it was difficult not to let slip something in conversations with his friends that could put them onto the path to the truth."[3] Such concerns about losing his priority might suggest that Lavoisier would have striven to keep his work to himself until he was ready to make it public. On the other hand, having protected himself by this action, he might have felt afterward less reason to guard his secret. In the longer memoir he drafted at about the same time, he placed his experiments among those rare events, capable of "making a revolution in science," which one should not "meditate in silence" until one had tied everything together, but should "hasten" to make known to the community of scientists.[4] Perrin has conjectured that Lavoisier did not present this paper, either because there was no time for him on the program, or because he changed his mind about the wisdom of a precipitous announcement.[5]

It is difficult to imagine that Lavoisier could have kept all knowledge of what he was doing private, especially when he began using the Tschirnhausen lens in the Louvre, in quarters belonging to the Academy. Whatever strategy he may have adopted to protect himself, it is not hard to imagine why he had become by March almost desperate to present his theory as auspiciously as he could manage, even in the face of his continued inability to tie his discoveries "together with other facts." He did not lack the bravado to do so.

Writing on the right side of each page of two folded sheets of paper, Lavoisier drafted for his presentation a memoir with the cumbersome title "On a new theory of the calcination and reduction of metallic substances, and on the cause of the augmentation of weight that they acquire by the fire, whether through calcination or through other analogous processes." He made some corrections between the lines of his text, then more extensive ones on the left side of each page, in some places crossing out whole paragraphs and replacing them with corresponding passages that he squeezed into the space he had left on the wide margins. He then made a clean copy of the corrected manuscript, and added further corrections on it. The successive versions of his memoir display significant developments, both in his interpretation of his investigation and theory, and in decisions about how to present them in the most favorable light. To describe all of the changes he made would require too much space for my present purposes. I cannot attempt here to present each of the stages in the evolution of the memoir as a whole, but will be content with discussion of the changes he made in critical paragraphs that illuminate most directly his struggle to represent his theory as strongly as possible in view of his experimental difficulties.

Lavoisier began his presentation with a dramatic flourish:

> The sort of fermentation that reigns in almost all of learned Europe about the nature and properties of fixed air, the great number of tracts that have appeared on this subject, [and] the great number of scientists who are occupied with it, do not permit me to postpone any longer to communicate to the Academy some experiments which I have deposited in its midst, and of which I was not intending to make use until I would have formed a complete body of theory.[6]

Recognizing afterward that the need to formulate a more complete theory had been a less urgent reason for his delay than the need to gather stronger experimental support for the theory that he had already in mind, Lavoiser rewrote the last three lines of his lengthy sentence, making it read that "some experiments that I had deposited almost a year ago in its midst, and which I had proposed to myself not to cause to appear until I had been entirely satisfied with my work."[7] The phrase "almost a year ago" that he inserted here is a suggestive slip. It had actually been only four and a half months since he had deposited with the secretary of the Academy the sealed note to which he referred. That this time seemed to him more like a year is a telling hint of the pressure under which he had felt himself since then to reach a position in which he could reveal what the note had hidden.

In a further alteration Lavoisier made in this paragraph, he acknowledged more openly the source of his anxiety. In place of "the great number of scientists," he put "the outstanding merit of some of those who are occupied" with the subject, and added "finally, the fear of seeing a discovery that I believe important taken away from French chemistry and from myself" had induced him to delay no longer making it public.[8] It is only when we have seen, through his laboratory notebooks, the extent of dissatisfaction that Lavoisier felt privately about his work at this time, that we can understand the reasons he gave for his action to be less an affirmation that he had met his earlier demands on himself, than veiled justifications for waiting no longer for that time to come.

"Present circumstances," Lavoisier continued in his first draft,

> do not permit me to give here the detail of my experiments, and I shall content myself with presenting the result. In our special sessions I shall return to each of these subjects [which he changed to "each subject in particular"], I shall give the details of the experiments [which he changed to "the operations"] the description of the different apparatus to which I was obliged to have recourse, as well as to the machines that I have employed.

Between the lines Lavoisier wrote, "I shall multiply the proofs as much as will be possible for me."[9] In the later revisions, however, he eliminated this phrase, and emphasized more strongly that it was the "Brevity of the time allotted to our public meetings [that] reduces me to presenting here only the principal results of my experiments."[10] The final version thus eliminated any suggestion, visible in the earlier ones, that his own circumstances, rather than the limitations imposed by the format of the public meeting, were causing him to postpone until later meetings the details of his experiments, and erased the hint that he might not yet be satisfied that he had sufficiently "multiplied" his proofs.

After mentioning the "remarkable" observations of the weight gained when metals are calcined, an effect that "scientists have not yet been able to explain in a satisfactory manner," Lavoisier came quickly to the heart of his experimental claim:

> If, instead of performing these experiments in the open air, one does them under a portion of air enclosed in a glass jar inverted in . . . a trough, and one intercepts the communication with the air of the atmosphere by means of water, oil, or mercury; to the degree that these metals are reduced to a calx the volume of the air diminishes, and the augmentation of the weight of the metal is found to be almost equal to the quantity of the air absorbed. If then, by means of a burning glass or some other procedure, of which I shall give the details, one is able to achieve the reduction of these metals, that is, to cause them to change from the state of the calx to that of the metal, they immediately restore all the air that they had absorbed and lose at the same time all the augmentation in weight that they had acquired.[11]

Placed in its temporal context this passage, to which Lavoisier made only stylistic revisions afterward, is breathtaking. In it we seem to see his vaunted balance sheet method suddenly emerge in its full glory. Knowing the actual state of his experimental progress at the time, however, we can also see how boldly he took advantage of circumstances that he claimed did not permit him to give the details of his experiments; for this was a wholly ideal situation that he portrayed, a summary of what he aspired to demonstrate, but what he was still far from achieving.

After mentioning that "other metals such as lead and tin and almost all semimetals" do not calcine easily, but require long exposure to a heat strong enough to melt them, Lavoisier described two of his recent experiments:

> I took **1 gros of lead and 1 gros of** [he then replaced the words highlighted here by "equal parts of lead and"] tin, which I mixed together and I kept them for a long time exposed to the focus of a burning lens **in a suitable vessel** [he then replaced the highlighted words with "under a glass jar in a given quantity of water"]. I succeeded in calcining them

> fairly well, but not as promptly as in free air. When the operation was finished, the quantity of air absorbed was found equal to 125 times the volume of the materials employed.

"The reduction of minium and litharge," Lavoisier continued,

> gave me the inverse of that experiment. One knows in general that every metallic reduction is accompanied by a movement of effervescence, that observation is found in all treatises of chemistry and metallurgy. I assured myself through a number of experiments that this effervescence is due to fixed air which is disengaged. I shall describe [later] the apparatus by means of which I succeeded in retaining it and measuring its quantity. It is especially on minium and litharge that I have multiplied these experiments. One reduces them easily by a simple addition of one-twelfth of their weight of powdered charcoal. This experiment has always given me from **2 or 4** hundred times its volume of air [he subsequently crossed out "2 or 4" and put instead "at least 3"], with some variation which derives from a circumstance of which I shall give an account in the future.

On the left margin Lavoisier tried twice to find ways to acknowledge further the experimental problems alluded to vaguely here, while withholding the details for another occasion. First he wrote, "I admit some exceptions to these rules that I have sometimes observed, which I shall explain in other circumstances." Preferring not to state this qualification in the form of an "admission," he crossed the statement out and wrote it in another way: "I have encountered exceptions to these general rules, which derive from particular circumstances that I shall explain at another time."[12]

In the next draft of the memoir, Lavoiser rewrote his account of these experiments in a more compact way:

> There is absorption of air during calcination, disengagement of air during reduction. Equal parts of lead and tin, exposed for a long time under a glass jar to the focus of the great burning lens ["gave me" Lavoisier wrote but then crossed out] absorbed, even though the calcination was not complete, 125 times their volume of air. On the other hand, minium, which is nothing but the calx of this same lead, mixed with one-twelfth [part] of powdered charcoal has constantly given me, whether through the fire of a furnace or the heat of the burning lens, a disengagement of air equal to at least 300 times the volume of the reduced lead. I have sometimes encountered **exceptions to these general rules**, [he crossed

> out the words highlighted here and replaced them with "differences in these results"], they regard particular circumstance about which I shall give an account at another time.

Finally Lavoiser eliminated the last sentence.[13] In these revisions of his descriptions of the experiments most critical to the general theory he wished to propose, we can see him struggling to find a way to acknowledge his experimental difficulties without undermining the general claims he was about to make. In the end he chose not to allow his audience even the glimpse of his problems that he had allowed for a time to enter his description. The description itself moved away from a brief narrative, in the first person, of what he had done, to an impersonal, third person, summary of a generic set of experiments of the type he had carried out. The final version betrayed no traces of the qualms that had so recently beset him. It would be hard to imagine a more guarded, more selective, or more optimistic presentation of the results of Lavoisier's so far tortured experimental efforts.

The thinness of his experimental foundations did not hinder Lavoisier from drawing far-reaching conclusions:

> From these experiments it evidently results 1) that a metal calx is nothing but the metal itself combined with fixed air, 2) that a metallic reduction consists in nothing but the disengagement of the air of metallic calces, whether presented to it by a body with which it has more analogy, or otherwise, and 3) that it is to the fixed air contained in the atmosphere that the metals owe the augmentation of weight.[14]

Evidently Lavoisier had by now completely repressed the suspicion aroused by his experiments with minium and volatile alkali that the air in the calx might not be fixed air, and had decided, publicly at least, to suppress the private doubts he still felt about his theoretical system as a whole.

With a boldness tempered by some signs of caution, Lavoisier seemed ready to launch the revolution in physics and chemistry he had contemplated, by a direct attack on the followers of Stahl. "This theory," he wrote,

> is destructive of that of Stahl adopted today by almost all chemists, and this circumstance would be appropriate to put me on guard. I have, nevertheless, not been able to reject the evidence, especially when decisive experiments assure me that it is possible to reduce almost all

> metals without the addition of phlogiston. I shall return to this subject on another occasion.

Prudence immediately prompted Lavoisier to restate his position in a somewhat less militant, but also less confident tone. In the left margin he wrote, "This doctrine is directly contrary to that established [Lavoisier then changed this last word to "taught"] by Stahl and adopted a generation ago by all chemists. That circumstance is capable of making me distrustful, and putting me on guard against my own experiments."[15] If the first statement had been too strong for him, this one was too weak. He drew a line through it and postponed his discussion of the topic to later in his draft.

The brevity imposed on his presentation to the Academy did not prevent Lavoisier from discussing a wide range of phenomena that he could interpret in terms of his new theory. Without mentioning his current inability to detonate niter, he explained the process as another means to supply air to metals. Drawing implicitly on the single experiment in which he had compared the precipitate formed by the actions of caustic and mild alkali on a dissolution of mercury in nitrous acid, he described the general action of alkalis on metallic dissolutions as a third way to join fixed air to metals. Generalizing boldly that "all metallic substances without exception are susceptible to combine with fixed air" by one means or another, he gave arguments for not excluding even gold from that claim. Despite "the restrictions" he had "prescribed for myself," he could not forebear to describe some experiments which "throw a great light on the nature of acids and the phenomena of fermentation." The absorption of air that he had demonstrated for sulfur and phosphorus was not unique to those substances, he confidently assured his audience, "it is the same in the formation of all acids." To support his bold induction he adduced the example of the fermentation of sugar, in the late stages of which the air earlier formed is reabsorbed and "enters into the composition of the acid" which forms then. These observations, which he apparently drew from a treatise by the Abbé Rozier, shed a "completely new light on the phenomena of fermentation, and put me in a position, as I hope, to give soon a nearly complete theory." That claim appearing too optimistic even for him, Lavoisier crossed it out.[16]

In the last part of his presentation, Lavoisier circled back to Stahl's phlogiston theory. "Just as the combustion of sulfur and phosphorus absorbs air," he wrote,

> so also every time one wishes to make sulfur with vitriolic acid there is an effervescence and disengagement of fixed air. That effervescence was announced by Stahl, and by all the chemists who repeated his experiment on the formation of sulfur.
>
> It follows from that, that the theory and the calculation by Stahl of the proportions of vitriolic acid and phlogiston that enter into the composition of sulfur are either entirely wrong, or at least susceptible to very considerable correction. According to the above experiments and observations, sulfur is nothing but very dry vitriolic acid deprived of air, united, no doubt with matter of fire and light. In the same way, vitriolic acid is a sulfur combined with water and air, and less charged, no doubt, with matter of fire. One could say as much of phosphorus and phosphoric acid.
>
> Finally, all the experiments that I have made on this subject lead me to believe that almost all of the phenomena that one attributes to phlogiston derive only from the absence of fixed air, and reciprocally, as I have shown for the metals.
>
> I have even come to the point of doubting if that which Stahl calls phlogiston exists, at least in the sense that he gives to that word, and it seems to me that in every case one could substitute the name of matter of fire, of light, and of heat.[17]

As if pondering, even as he wrote these lines, whether he could so abruptly dismiss a theory held by "all chemists," Lavoisier finished his draft by stating that

> all of this ought, naturally, to be preceded by a historic summary of the successive discoveries that have been made on fixed air and [then] that which I have added would become much more intelligible. Circumstances have imposed that transposition on me. I shall reestablish everything in its order in our particular meetings.[18]

In his final draft, Lavoisier eliminated this long discussion, along with the statements earlier in his paper about the opposition between his theory and that of Stahl. In place of the latter he wrote, "I will not examine at all, at this point, whether the phlogiston of Stahl takes any part in the phenomena of the calcination and reduction of metals. I shall treat these matters in a separate paper, and I shall have, in that regard, very remarkable things to announce to the Academy."[19] In place of the closing discussion of the first draft he wrote,

> Here would be, no doubt, the place to draw some reflections on the doctrine of Stahl, on the principle that he calls phlogiston, . . . and the name of which chemists since him have often abused. One knows, in fact, that it is to the absence or presence of that principle that he attributes calcination and reduction of metals, the transformation of sulfur into vitriolic acid, of vitriolic acid into sulfur, and a great many chemical phenomena. But my experiments are not yet complete enough to dare to challenge that celebrated chemist.[20]

Why did Lavoisier back away from the challenge he had seemed, in his first draft, ready to take up? Perrin, who first drew attention to this shift, wrote simply that "the criticism of Stahl was muted."[21] Instead, Lavoiser made a far more global prediction. Avoiding the specific target to which his earlier draft had been directed, in a paragraph following the above, he asserted that he had said enough to show that "the present theory of chemists is defective on many points, and it is probable that the more profound phenomena of fixed air will lead this science into the epoch of an almost complete revolution."[22]

Several years later, when he *was* ready to confront the doctrine of Stahl publicly, Lavoisier recalled that he had been "stopped" from doing so earlier because he had too "little confidence in his own enlightenment" to dare put forth "an opinion that might appear peculiar and that was directly opposed to the theory of Stahl and of many celebrated men who followed him."[23] Was it simply timidity, particularly the fear of confronting, in the Academy itself, some of those celebrated followers of Stahl, that induced Lavoisier to dilute and to defer the direct challenge to specific manifestations of Stahl's theory that he had contemplated when he wrote his first draft? Were his vaguer promises of revelations to come, and intimations of a complete revolution ahead merely maneuvers of someone suddenly afraid to open the contest he foresaw? Was the potential presence in his audience of such eminent Stahlian chemists as Pierre Joseph Macquer enough to make the younger chemist flinch?

In the light of the extremely incomplete state in which we have seen Lavoisier's experiments to have been, and in view of the objections that we see he himself saw to his theory that he could not resolve, we can view his change of heart in a different way. Perhaps he was still young enough to be awed at the prospect of actually launching the total revolution he foresaw, but he must also have recognized, in the very act of composing his attack, that he was not yet ready to carry it out.

In the ending that he substituted in the final version of his paper for the criticism of Stahl in the earlier version, Lavoisier added a tribute to a valued councillor. "In terminating this memoir," he wrote,

> I owe to the public the acknowledgment that the initial ideas for the theory that it contains do not belong to me. I have drawn them from conversations with a distinguished academician who divides his time between the functions of higher administration and the study of all the sciences, and who holds grand views on every subject with which he occupies himself. It is almost unnecessary that, to these traits I add the name of M. de Trudaine.[24]

Historians have long recognized the association of Lavoisier with Trudaine de Montigny, director of the *Bureau de commerce* of the French government, honorary member of the Academy of Sciences, and a keen amateur scientist who had equipped his own chemistry laboratory. Guerlac describes Trudaine as "a friend, patron, disciple, and (for a brief time and modest extent) the collaborator of Lavoisier."[25] If we take Lavoisier's acknowledgment literally, Trudaine contributed far more to the development of Lavoisier's "new theory" than Guerlac's characterization of their interaction implies. Unfortunately, except for a letter written in the summer of 1772 urging Lavoisier to examine a procedure that Joseph Priestley had published for saturating water with fixed air,[26] there is little surviving record of the specific nature of Trudaine's contributions to the earliest development of Lavoisier's thoughts and experiments. The insertion of the above tribute to Trudaine in the paragraphs with which Lavoisier replaced his criticism of Stahl suggests, however, the possibility that one of the good ideas Trudaine gave him was to postpone that move until he could strengthen his own position.

At the public assembly of the Academy held on April 21, five lectures filled the meeting. The first two were a report on a voyage to the Antilles and a eulogy of a distinguished foreign associate of the Academy. Lavoisier gave the third talk. The title, as reported in the *Mercure de France*—"Memoir on the effects and the action of fixed air on the calcinations and reductions of metallic substances"—differed somewhat from that of the manuscript he presumably read there, but clearly represented the same topic. His lecture was followed by one on the centering and decentering of arches in bridges, and one on the state of astronomy in the Indies.[27] It is notable that Lavoisier was the only person at this semiannual display of the activities of the Academy to the public, to present a theoretical contribution to a science.

Even without a direct assault on the doctrine of Stahl, The memoir with which Lavoisier announced to the world his "epochal"[28] new theory remains an astonishing performance. Having failed to produce the experimental support that he himself had believed he needed, having failed to resolve a major contradiction that he himself perceived in his theoretical structure, having run out of time to emerge from his difficulties, he prepared to present his innovation as solidly grounded and as the harbinger of a revolutionary epoch in chemistry. When Lavoisier's public pronouncements at the Easter meeting of the Academy are examined in the light of his private experimental tribulations of the time, he appears so unlike the perspicacious, methodical, sure-footed young scientist that we have been accustomed to see, so precarious in his prediction of the revolution that we have taken to be the mark of his prescience, that we may be tempted to think that we have exposed instead an agile posturer. That would be a mistake. That Lavoisier possessed from the beginning of his scientific activity vaulting ambition and hunger for early recognition has long been obvious. The brashness with which he brushed aside, in the august setting in which he announced the theory with which he hoped to shake the foundations of physics and chemistry, the obstacles that he still confronted in the laboratory, is just another example of these youthful traits. What more than compensated for such foibles was that Lavoisier possessed the drive and the resourcefulness not to be deterred from his course. Like other gifted experimentalists, he probably possessed also the special intuition that enabled him to sense, long before he could know, that the obstacles in his path were merely technical difficulties that he could overcome. Finally, he had the good judgment never to publish the paper that he read at that Easter meeting.

✻ CHAPTER FIVE ✻

Reflections

EVEN before he presented his paper at the Academy, Lavoisier wrote, on April 15, an "Essay on the Nature of Air," which he intended to be a sequel to the first. In it he discussed some ideas about the three states of matter whose origin can be traced back, as J. B. Gough first pointed out in 1968, to a manuscript that the youthful Lavoisier had written in 1766. Then, after stating that when water has acquired a certain degree of heat it becomes a vapor, Lavoisier had asked himself, "Is air itself not a fluid in expansion?"[1] In August 1772, he had returned briefly to the same idea at the end of a long memorandum on air, claiming as his "completely new opinion," that "the air we breathe is not a simple entity, it is a particular fluid combined with the matter of fire."[2] In April 1773, he developed for the first time the broader implications of this novel viewpoint. "All natural bodies," he now asserted, "appear to our eyes in three different states." Some are solid, some fluid, some in a state of expansion that he named the "vapor" state. The heart of his novel view was that "the same body can pass successively through all these states, and for this phenomenon to take place requires only that it combine with a larger or smaller quantity of the matter of fire." These changes of state apply, he stated confidently, "to all natural bodies without exception." From the fact that water, mercury, ether, and other fluids can pass into a vapor state when heat is added to them, should we not conclude, he asked, that "the air itself is nothing but a fluid in expansion?" To dramatize his view, he asked his readers to transport themselves to a planet much warmer than the earth. There, substances that are solid here would be fluids, those that are fluid here would be vapors there. On a planet at the extremes of the solar system, "for example, in the region of Saturn," on the other hand, the opposite effects would occur.[3]

Lavoisier claimed that these conclusions were "proven by decisive experiments," but gave in support of them only general observations concerning heat absorbed or released in various chemical operations. The most impressive argument he made, that of the imagined situation on other planets, was a thought experiment, rather than one conducted in his laboratory. The new viewpoint he had expressed here was germinal to one of the most profound conceptual changes in the history of chemistry, but at this stage Lavoisier had yet to make a strong connection between it and the long experimental program on which he had embarked.

It is a mark of the fluidity of the theoretical structure he was striving to develop, that at almost the same time that he had, in the first draft of his memoir on calcination, come to doubt the existence of phlogiston, he invoked in this memoir the word phlogiston over a dozen times as a real entity. It was, however, not exactly Stahl's phlogiston. In any process in which air is fixed, Lavoisier asserted, "there is a disengagement of phlogiston or matter of fire." In the reduction of metallic calxes, on the other hand, "it is the fixed air to which the phlogiston is joined, not the metal. It is not that I absolutely deny that some metals may contain a little phlogiston, my experiments have, in that regard, not yet led to decisive results."[4] The comparison of these passages with those in the draft of the other text in which Lavoisier described his experiments as destructive of phlogiston, reveal a vacillation between two viewpoints that coexisted in his mental world for longer than most historians have recognized. One view, conditioned by nearly a decade of learning and practicing a chemistry in which phlogiston was as familiar to him as the reagents he used, was to modify the properties of phlogiston to harmonize it with his experiments on combustion and calcination, and to identify it with his own emerging concept of the role of fire in the three states of matter. The other view, perhaps reflecting his ambition to bring about revolutionary changes in chemistry and physics, was to challenge phlogiston as no more than an outmoded "doctrine."

During the same weeks that he launched his experimental program on the processes that fix and release airs, Lavoisier undertook a complementary literary project. Having immersed himself in the earlier literature on the subject, he saw that "a great number of foreign physicists," especially in England, Germany, and Holland, were engaged in research on it, but that French works on chemistry, even the "most modern and complete ones," were almost entirely silent on the topic. He decided that he could perform a service to his compatriots, in the "simple role of a historian," by preparing a summary of all that had been written about "the elastic emanations that are disengaged from bodies during combustion, during fermentation, and during effervescences."[5] Not content to wait for the publication of his essay to make known what he had found, he read part of it at the regular meeting of the Academy on April 30, and continued to read from it on the 19th and 28th of May.[6]

Lavoisier covered his subject with extraordinary thoroughness, making long summaries of works beginning with Paracelsus and Van Helmont, and ending with a memoir that had been read by his fellow French chemist Jean-Baptiste Bucquet at the Academy just one week before Lavoisier himself began the reading of his history.[7] The extent to which he enlightened his colleagues there is not known, but what is evident from the contents (as published nine months later in his *Opuscules physiques et chymiques*) is the degree to which Lavoisier deepened his own knowledge of the subject through his efforts to summarize clearly and succinctly what his predecessors had achieved. Particularly important to the development of his own views were his treatments of Stephen Hales, Joseph Black, and Joseph Priestley.

As Guerlac has stressed, Hales was the only one of these three with whose work Lavoisier was familiar before the fall of 1772.[8] Nevertheless, his careful reading of the long chapter on the analysis of air in Hales *Vegetable Staticks* in preparation for his historical essay also conditioned the evolution of Lavoisier's views during the early weeks of the research program he began in February 1773. What most impressed him about that work? The experiments of Van Helmont and of Boyle, Lavoisier wrote, had already shown that in many operations a large quantity of an elastic fluid analogous to air is disengaged or absorbed, but

> one had no idea yet of the quantities produced or absorbed. M. Hales was the first to envisage the subject from that point of view. He devised various simple and convenient means to measure the volume of the air with exactitude.

Four of the fourteen pages that Lavoisier devoted to his summary of Hales's experiments consisted of tables in which he had gathered together their quantitative results. Lavoisier represented the results, as Hales had originally described them more diffusely in his text, as comparisons between the volume of the air and of the body that had absorbed or released it. (In a few cases he gave also the weight of the latter, but in the tables as a whole the volume relations are predominant.) In the remainder of his discussion, Lavoisier picked out various qualitative observations, such as that "the acids in general, and spirit of niter in particular, contain air," which were pertinent to his own interests.[9] It is difficult to sort out which of these observations might have

EXPÉRIENCES PAR LA DISTILLATION.

NOMS DES MATIERES mises en expérience.	NOMBRE de pouces cubiques d'air, produits par la distillation.
Sur les Végétaux.	
Un pouce cubique ou 270 grains de bois de chêne..............................	256
Un pouce cubique ou 398 grains de pois....	396
142 grains de tabac sec....................	153
Un pouce cubique d'huile d'anis...........	22
Un pouce cubique d'huile d'olive...........	80

NOMS DES MATIERES mises en expérience.	NOMBRE de pouces cubiques d'air, produits par la distillation.
Un pouce cubique de tartre...............	504
Un pouce cubique ou 270 grains d'ambre..	270
Sur les Substances Animales.	
Un pouce cubique de sang de cochon, distillé jusqu'à siccité........................	33
Un peu moins d'un pouce cubique de suif...	18
Un pouce cubique ou 482 grains de pointes de cornes de daim....................	234
Un pouce cubique ou 532 grains d'écaille d'huitres..........................	324
Un pouce cubique de miel................	144
Un pouce cubique ou 253 grains de cire jaune...............................	54
Une pierre de vessie humaine de $\frac{3}{4}$ de pouces cubes du poids de 230 grains...........	516
Sur les Minéraux.	
Un pouce cubique ou 316 grains de charbon de terre.........................	360 *
Un pouce cubique de terre franche.........	43
Un pouce cubique d'antimoine............	28

* C'est environ 102 grains d'air, suivant M. Hales; c'est-à-dire, le tiers du poids total.

NOMS DES MATIERES mises en expérience.	NOMBRE de pouces cubiques d'air, produits par la distillation.
Un demi-pouce de sel marin, & un demi-pouce d'os calcinés..................	64
Un demi-pouce cubique ou 211 grains de nitre avec de la chaux d'os calcinés.......	90

EXPÉRIENCES SUR LA FERMENTATION.

42 pouces de petite bierre en sept jours.....	639
26 pouces cubiques de pommes écrasées en treize jours........................	968

EXPÉRIENCES SUR LES DISSOLUTIONS ET LES COMBINAISONS.

NOMS DES MATIERES mises en expérience.	NOMBRE de pouces cubiques d'air, produits.	NOMBRE de pouces cubiques d'air, absorbés.
Un demi-pouce cubique de sel ammoniac avec un pouce cubique d'huile de vitriol, le premier jour......................	5 à 6	
Les jours suivans, il y en eut quinze d'absorbés.		
Six pouces cubiques d'écailles d'huitres, & autant de vinaigre distillé en quelques heures.....	29	
En neuf jours, il s'en est détruit 21, & les 8 autres disparurent en jettant de l'eau tiéde sur le mélange.		
Deux pouces cubiques d'eau régale versés sur un anneau d'or applati....................	4	
Deux pouces cubiques d'eau régale versés sur $\frac{1}{4}$ de pouce d'antimoine, en trois ou quatre heures....................	38	

16 PRÉCIS HISTORIQUE

NOMS DES MATIERES miſes en expérience.	NOMBRE de pouces cubiques d'air, produits.	NOMBRE de pouces cubiques d'air, abſorbés.
Quelques heures après, il s'en trouva 14 de détruits.		
Un pouce cubique d'eau-forte verſé ſur un quart de pouce d'antimoine en pluſieurs fois...	130	
Un pouce cubique d'eau-forte ſur un quart de pouce de limaille de fer..................	43	
Un quart de pouce de limaille de fer, & un pouce cubique de ſoufre en poudre..........	....	19
Un pouce cubique d'eau-forte verſé ſur autant de marcaſſite en poudre..................	....	85
Un pouce cubique d'eau-forte ſur autant de charbon de terre, 18 pouces, dont 12 furent reproduits les jours ſuivans.....	....	18
Deux pouces cubiques de chaux vive, & quatre de vinaigre...	....	22
Deux pouces cubiques de chaux, & autant de ſel ammoniac....	....	115
De la charpie trempée dans du ſoufre fondu, enflammée, abſorba dans un grand vaiſſeau..	....	198
Dans un vaiſſeau plus petit.....	....	150

NOMS

SUR LES ÉMANATIONS ÉLASTIQUES. 17

NOMS DES MATIERES miſes en expérience.	NOMBRE de pouces cubiques d'air, produits.	NOMBRE de pouces cubiques d'air, abſorbés.
Deux grains de phoſphore de Kunkel..................	....	28
Après l'inflammation, il n'avoit perdu qu'un demi-grain; quelque temps après, ſon poids ſe trouvoit augmenté d'un grain.		
Un morceau de papier brun trempé dans une forte ſolution de nître, & enflammé ſous une cloche par le moyen d'un verre ardent, produiſit..........	80	
En quelques jours, cette quantité d'air diminua.		

EXPÉRIENCES SUR LES CORPS ENFLAMMÉS ET SUR LA RESPIRATION DES ANIMAUX.

Une chandelle allumée, de $\frac{3}{5}$ de pouces anglois de diametre....................	....	78

B

3. Tables compiled by Lavoisier to summarize measurements of air disengaged or absorbed in experiments by Stephen Hales. Reproduced from *Opuscules*, pp. 14–17.

been germinal to his thought, and which ones only supported directions of thought that he had already taken up on other grounds, because of the lack of evidence about whether he had previously read Hales's work in detail, or knew of it before only indirectly through such sources as the chemical lectures of Guillaume Rouelle.[10] What seems clear is that his admiration for Hales's "exact" quantitative results reinforced the quantitative aspect of his own emerging experimental program, but also tended to concentrate his attention on the volumes rather than the weights.

Lavoisier probably read the works of Joseph Black and Joseph Priestley for the first time in the French translations of their publications which appeared during the spring of 1773 in the journal *Observations sur la Physique*.[11] The effects of both authors on him were profound. He entitled his summary of the former, "Theory of Black on the fix or fixed air contained in calcareous earths, and on the phenomena which the privation of this air produces in them." After noting the reasons for which Black had applied Hales's term, "fixed air" to a "species of air different from common elastic air," Lavoisier gave a general qualitative summary of the experiments through which Black had identified fixed air, its transfers from alkalis to calcareous earths, and its removal from these substances when they are combined with acids. He did not include any of the measurements Black had given of the weights gained or lost in these operations, but he did remark that when lime is precipitated from an acid solution by a caustic alkali,

> what is most remarkable is that the limestone loses, according to M. Black, the same quantity of its weight in that experiment as it does by calcination, and that it recovers its original weight when one precipitates it in the form of calcareous earth, that is, with all of its air.[12]

The most important impact of his study of Black's experiments on Lavoisier was probably that it prompted him to try to repeat the experiments using Hales's apparatus to measure directly the fixed air released or absorbed.

Lavoisier devoted far more space to Priestley than to any other of the authors he reviewed. This work he praised as "the most laborious and most interesting that had appeared since Hales," and the best suited to reveal "how much physics and chemistry still offer new routes to follow."[13] Among the horizons that Priestley opened for Lavoisier was the realization that there were distinct species of air beyond ordinary and fixed air. Priestley described experiments with the inflammable air discovered by Henry Cavendish, as well as his own discovery of acid air and nitrous air. He presented his method of testing the respirability of air through the "nitrous air test," instead of directly on animals, and he showed that plants restore atmospheric air "vitiated" by the respiration of animals.[14] The experiments of Priestley had not only "pushed back the boundaries of our knowledge of subjects" such as fixed air, "but one owes to him some facts which seem to discover a new order of things."[15] All of these new dimensions eventually expanded the scope of Lavoisier's investigative program immensely. It was, however, mainly in the second year of his progress that he began to absorb them into his research pathway. Therefore, a detailed discussion of Priestley's influence on him lies beyond the scope of this volume.

* CHAPTER SIX *

In the Shadow of Black

AFTER making his theory of calcination and combustion public at the Easter meeting, Lavoisier was free to pursue his research program in a somewhat less stressful manner. For the moment he put aside the frantic effort to make the calcination experiments work, and explored some of the other questions that had arisen both in his own investigation and in his study of the work of others. He also had time again to indulge in his penchant for designing special apparatus suitable to his needs. Probably the first step he took (although it is not recorded in his laboratory register), was to arrange to have crafted a retort able to withstand the extreme temperatures to which his experiments would expose it without cracking or melting as the conventional clay and glass ones had done. He decided to have it made of iron. M. Delorme, a craftsman residing on the rue de Charonne, in the faubourg S. Antoine, was able to produce for him a retort, shaped like an ordinary one, but fabricated from four pieces of forged iron: one shaped like a skull cap to form the base, and three collar-shaped ones. The edges of the pieces had to be adjusted to fit each other exactly. Then Delorme soldered the pieces together with copper.[1]

While waiting for his retorts to be ready, Lavoisier turned to types of experiments that would not require them. On May 4, he tried to reproduce, in a closed vessel, the fundamental Stahlian experiment of the reduction of vitriolic acid to sulfur. Placing a few pieces of charcoal and some vitriolic acid in a little glass capsule under one of his inverted jars, he kept it in the focus of the burning lens for several separate periods of time totaling a "good hour." When the focus fell on a piece of charcoal "imbibed" with the acid, the acid boiled, and white vapors circulated through the jar. It cleared after he stopped heating, and there was "a slight lowering of the water in that operation, but less than I had thought."[2] By analogy with the reduction of lead ore, he would have anticipated a substantial release of air. As so often before, Lavoisier again found it more difficult than he expected to confirm experimentally the inferences he drew from his theoretical views.

Lavoisier examined next another form of combustion, the burning of spirit of wine. He attempted to ignite the inflammable fluid under the jar by placing a piece of phosphorus in it, and focusing the burning lens on the phosphorus. The phosphorus melted, "but I could absolutely not enflame it, because, resting always at the bottom, it could not come into contact with the air necessary to burn it."[3] Unforeseen pitfalls seemed to lurk at every step in his experimental pathway.[4]

At the meeting of the Academy of Sciences on May 5, the sealed note that Lavoisier had deposited on November 1, 1772, was opened, at his request, in the presence of the Academicians. His colleagues now learned that on that date he had written that "I have discovered around eight days ago that sulfur, on burning, far from losing weight, on the contrary, gains it." It was the same with phosphorus. The gain in weight came from a prodigious quantity of air fixed during the combustion. He had then reported briefly his first experiment on the reduction of litharge which confirmed his "conjecture," and which appeared to him one of the most interesting discoveries since Stahl.[5]

The purpose of a sealed note was to allow the author to have it opened if, before he was ready to make a discovery public, someone else announced a similar advance. Why did Lavoisier request that the note be opened two weeks *after* he had already presented his new theory to the Academy? The Secretary of the Academy recorded only that "the author has asked the present mention [of it] to preserve for himself the date." It is possible to imagine that someone had raised a question about his priority that made Lavoisier feel he must prove that he had made the initial discoveries five months earlier. Another possible explanation is that his recognition of how difficult it was to produce a "complete theory" had led him to see the advantage of making public the starting point for the investigation whose completion now seemed further away than he could imagine before encountering a series of experimental setbacks. There is no evidence to support an elaboration of either of these conjectures.

"The precipitation of metals dissolved in menstrua," Lavoisier had declared in the memoir announcing his new theory to the Academy in April, "furnishes still another means to unite them to fixed air, and consequently to reduce them to calxes."[6] On May 6, he began to explore this claim by dissolving a metal in an acid. In a conventional distillation vessel he placed exactly one pound of a preparation of spirit of niter (nitrous acid as it was also known in the eighteenth century) and slightly more than 1 pound of water, then gradually added 2 ounces, 1 gros, 36 grains of iron filings. The resulting lively effervescence produced heat and reddish vapors, but as far as he could tell, not very much air. After he had reached, or slightly overreached, the saturation point, he weighed the material remaining in the vessel again, and found a loss of 4 gros, 19 grains.[7] This was the first experiment, recorded in Lavoisier's notebook, in which he wrote down the weights of all the materials before and after the operation in the balance sheet form so familiar in his later publications. The loss of weight did not support his position, however, and he did not pursue the question further for now.

Again Lavoisier suddenly changed direction. Inspired by a description of an apparatus consisting of two bottles connected by a tube provided with a stopcock that Jean-Baptiste Bucquet had given in the paper that Lavoisier summarized in his historical essay, and that Bucquet used to generate fixed air which he could then pass directly into the other bottle, where it acted on various substances, Lavoiser devised a modified version which he described in his notebook on May 7 as a "means to obtain fixed air without a bladder by an apparatus which leaves no doubt about its purity." He sketched the apparatus on the page facing its description (see fig. 4). In the bottle on the right, which he called, following Bucquet, the "mixing bottle," he placed chalk. He filled the funnel with vitriolic acid, whose flow into the bottle he controlled by means of a rod with a clay bulb at its tip that he inserted into the funnel until its tip blocked the opening, then raised to allow the acid to enter a little at a time. The fixed air generated by the action of the acid on the chalk passed through the tube and stopcock into the "receiving bottle" at the left, which was an inverted jar filled with water and immersed in a water bath. As the air collected in the jar, it displaced the water in the usual fashion of a pneumatic trough. Lavoisier thought that the apparatus could serve also for other mixtures, and, with precautions taken to avoid corrosion, for metallic dissolutions and for producing Priestley's nitrous air. To conserve the air after the bottle had been filled, Lavoisier devised a method to introduce a layer of oil on top of the water and to stopper the bottle.[8]

A similar apparatus (fig. 5) intended for an "experiment on the effect of very deep cooling on fixed air" was also stimulated by the reading Lavoisier had done for his historical review. The Count Saluces of Turin had reported that fixed air allowed to stand for twelve hours at the temperature of freezing water acquired the properties of ordinary air. Lavoisier planned to test the Count's claims with an apparatus which permitted the fixed air to pass from a bottle packed in ice to a receiving bottle like the one in the other apparatus.[9]

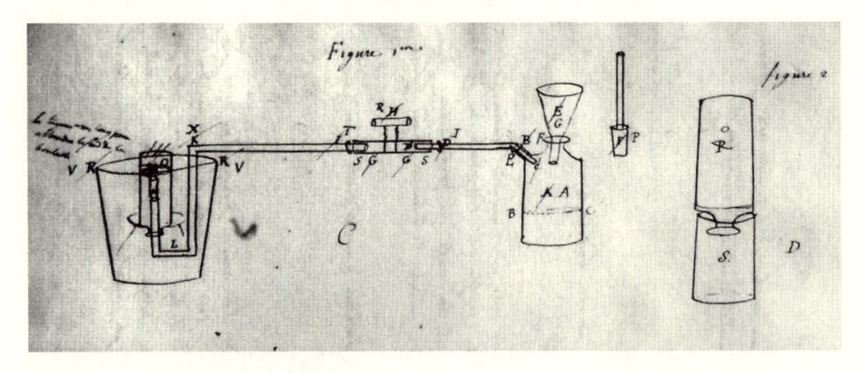

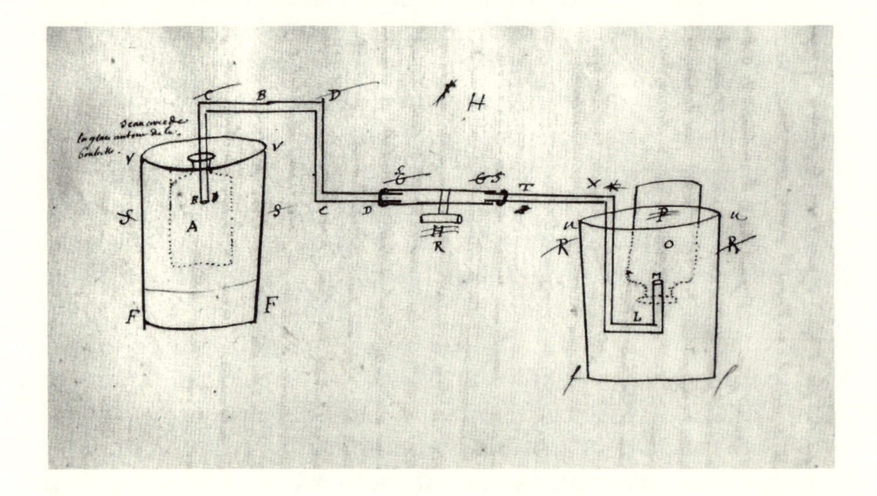

The third apparatus Lavoisier designed was intended to "cause any air one wished to submit to the examination to pass through lime water, or through any other liquid one wishes." This somewhat more complicated setup included an inverted jar containing the air to be tested, connected via a tube with a stopcock to an air pump used to draw it out of the bottle and pass it successively through three bottles containing limewater. The tubes inserted into each bottle forced the air to the bottom, where it emerged and bubbled up through the liquid before passing into the next tube. Whatever air was not absorbed in these passages collected at the other end of the apparatus in a second pneumatic jar (see fig. 6).[10]

Unlike the apparatus Lavoisier had designed two months earlier, these required only the assembly of readily available components and

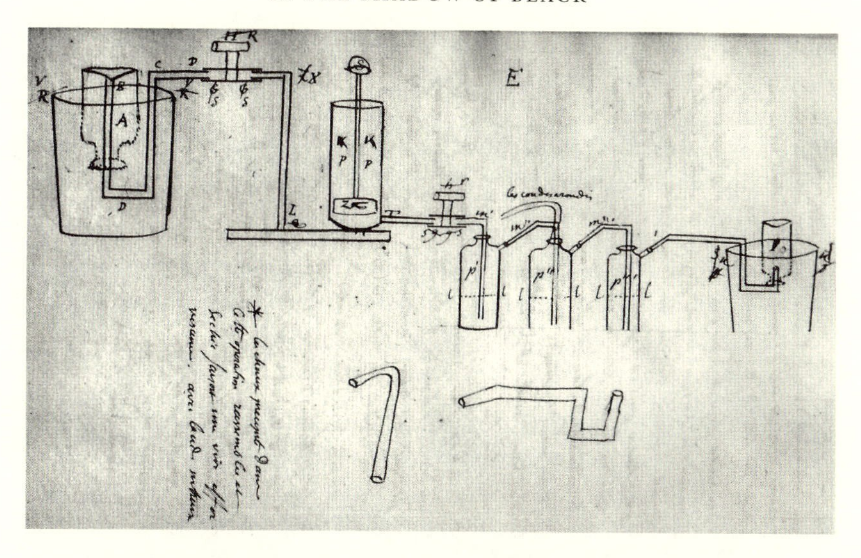

4–6. Sketches drawn by Lavoisier in his laboratory notebook of apparatus to be constructed. For descriptions of purpose of each, see text. Reproduced from Lavoisier, Cahier lab. 1, facing pp. 33, 35, 37.

careful luting together to avoid air leaks. Lavoisier quickly had them ready for use. In his first experiment with them he combined the second and third apparatus to pass fixed air refrigerated to 15 degrees below freezing, by packing it with a mixture of salt and ice, through the three lime water bottles. A defect in a lute blocked the passage of the air into the second bottle, but the fact that the first bottle became turbid was already evidence enough to doubt that the fixed air had become ordinary air. As a control, Lavoisier bubbled common air through one of the lime water bottles. It passed through without causing any turbidity. Lavoisier verified that the precipitate from the fixed air was calcareous earth by causing it to effervesce with nitrous acid.[11] The new apparatus was already proving effective. Soon it became essential to Lavoisier's entire research program.

To subject the Count Saluces's idiosyncratic claim to experimental test was only a subsidiary motivation for the construction of the new apparatus. Lavoisier's main reason was probably that his study of the history of the release of elastic fluids had convinced him that a high priority should be to repeat Joseph Black's experiments, but employing new methods to collect and measure the air that Black had detected only indirectly through the gains and losses of weight of the substances which combined with it or were deprived of it. Before beginning those experiments, however, he paused to do a few experiments using the Academy's "pneumatic machine" (vacuum pump). He was still vexed by the problem posed to his theory by the relation of minium and volatile alkali. On May 10 he again expressed his concern in his notebook:

> Volatile alkali extracted from sal ammoniac by minium is truly a caustic alkali. I saturated it with lime water and it formed no precipitate at all. This fact is extremely peculiar and appears to me entirely inexplicable according to the principle of the fixation of air in metallic calces.

Placing liquid volatile alkali prepared with minium under the pneumatic machine, he began to pump. When the barometer descended to 10 inches the fluid began to bubble, and the bubbling was "well established" by the time the pressure reached 7 inches. Strings of bubbles arose in considerable quantity from the bottom of the containing vessel, but Lavoisier was surprised to see that they did not cause the barometer to rise. Instead it descended further, becoming stationary at 6 1/4 inches. The ebullition ceased. With another stoke of the pump he lowered the pressure to 5 1/4 inches, but there was only a brief further ebullition. Applying himself vigorously to the pump, he got the barometer down to 1 1/2 inches. At each stroke there was a brief ebullition. Although he continued to work the pump, he could lower the pressure no more, nor excite any more ebullition. He repeated the experiment with a thermometer, and recorded the fall in temperature accompanying the fall in pressure. The results were similar.[12]

There is no indication that this strenuous effort helped Lavoisier to resolve his anomaly. Like other creative investigators, he went on with the other aspects of his agenda, undoubtedly not hiding from himself the seriousness of the difficulty, but trusting that future developments would lead him to the explanation that was, for now, beyond his reach.

On the same day, Lavoisier heated spirit of wine to 30 degrees and placed it under the pneumatic machine, "believing," he noted, "that it would boil easily." Once again, the result defied his expectations:

> At the end I had a few puffs of vapor which caused the mercury column to vary, but I could not be sure whether there was a rise or a fall. I repeated the experiment with spirit of wine heated to the boiling point. It did not appear to boil sensibly more quickly under the recipient. I also had a puff of boiling, after which the liquid remained still. I repeated the

> experiment several times, and it appeared to me constantly that the spirit of wine did not boil more rapidly than water, [but] that it supports a shorter column of mercury.
>
> I believe that I could just see also that it boils as easily cold as warm under the pneumatic machine.

Lavoisier next tried turpentine. Under the recipient it threw off small bubbles when the air had been exhausted, but "it did not appear that they caused the mercury to rise."[13] All of his experiments on these two volatile liquids thus proved indecisive. What did Lavoisier have in mind in carrying them out? Most likely they represent his first effort to test the ideas he had developed in April concerning the three states of matter. If the liquids had boiled as rapidly as he had expected, and caused the mercury in the barometer which measured the pressure in the pneumatic machine to rise, that would have been a strong indication that he had produced and contained spirit of wine and turpentine in the vapor state. As matters turned out, it was only an initial exploration of the question. For the moment, he went no further. Three days later he began preparations to repeat the experiments of Joseph Black.

On May 13, Lavoisier took 12 ounces of "very regular" crystalline needles of niter, containing "no vestige of water of crystallization or marine salt," and melted it in a crucible for half an hour. Afterward it weighed 11 ounces, 2 gros, 32 grains. He reduced it to a powder and filtered it through a silk sieve and then through one made with horse hair. His purpose was evidently to procure a supply of very pure, desiccated niter to use in experiments involving the action of nitrous acid on chalk. In preparing desiccated chalk he again encountered the unexpected. After drying 5 ounces of finely powdered chalk previously passed through a silk sieve, in an iron ladle for half an hour, while keeping the heat low enough to avoid reddening it, he found that it now weighed 4 ounces, 7 gros, 46 grains. "That appeared very extraordinary to me," he remarked, "because the lime had been very humid before this operation," and ought therefore to have lost more than 26 grains of water. "May it be," he asked, "that lime calcined by a mild fire recaptures fixed air?"[14] According to Black, "when calcareous earths are exposed to the action of a violent fire, and are thereby converted into quicklime, they suffer no other change than the loss of a small quantity of water and of their fixed air."[15] By heating his chalk more gently, Lavoisier had intended to remove the water without removing the fixed air, but the weight loss was so small that he apparently suspected that the substance had actually gained fixed air in the process. There is no indication whether he considered that possibility to pose a serious anomaly, or a secondary technical problem.

Before proceeding to the experiments for which he planned to use the desiccated niter and chalk, Lavoisier set up several fermentation experiments. The observation he had included in his Easter memoir, that the "abundant" air released in spirituous fermentation begins to be reabsorbed in the later stage, as the fermentation liquor becomes acid, an "experiment which throws a wholly new light on the phenomena of fermentation," Lavoisier apparently took from a treatise published in the fall of 1772 by the Abbé Rozier.[16] Now wishing to study these phenomena for himself, Lavoisier put 5 ounces of red wine into a medicinal vial on May 15, and placed it under a glass jar inverted, in the usual way, over water covered with a layer of oil. He evidently hoped that the wine would spontaneously turn acidic. On the next day he placed 1 ounce of wheat starch and 5 ounces of water into a similar vial, put another such mixture into a third bottle, and placed both vessels under a glass jar "with an apparatus suitable to measure the air." On the first day "elastic fluid did not appear to be disengaged," but it began to disengage "sensibly" on the following day. In a third vial in which he had placed 1 1/2 ounces of bran with a little over 5 ounces of water, there was "considerable production of air at the end of two days."[17] While these processes continued, Lavoisier undertook the experiments with chalk, lime, and nitrous acid for which he had been preparing.

"A calcareous earth deprived of its air, or in the state of quicklime," Black had written, "greedily absorbs a considerable quantity of water, becomes soluble in that fluid, and it is then said to be slaked."[18] Lavoisier began his repetition of Black's observations by measuring the quantity of water absorbed in the "extinction of quicklime." In an iron cauldron he put 32 ounces of quicklime and slaked it with water. Afterward three small pieces of stone, weighing 2 ounces, 2 gros, remained undissolved, so that he had "actually employed only 28 ounces, 6 gros of quicklime." Immediately placing the cauldron on the fire, he heated it for 7 to 8 hours, stirring and dividing the pulpy mass until it had dried to a white powder "which weighed, everything evaluated, just 37 ounces." Actually, he admitted, the measured weight was 1 1/2 gros less, but he had added a little for what "remained in the cauldron, on the spatula, etc." If he had actually employed 32 ounces, or 256 gros of pure quicklime, he calculated, he would have obtained 329 gros, or 41 ounces, 1 gros of slaked lime. "The proportion of quicklime to lime slaked and dried is, therefore, 1,000 to 1,287, or, to put it in another way, quicklime absorbs a little more than one quarter [of its weight] of water."[19]

If this experiment avoided all of the pitfalls that had plagued Lavoisier's efforts up until then, that was in part because it was extremely simple. The process was well known, it required no special apparatus, and no measurements of elastic fluids. The determination of the quantity of weight gained by subtracting the weight of the substance which entered the operation from the product imitated Black's approach to these phenomena. Lavoisier sought no innovative view of the process. Nevertheless, it marks a significant step in the evolution of his experimental style. Undoubtedly he made his weighings as accurately as he could, but he was nevertheless willing to correct one of the two measured values by about 4 percent by guessing at residual amounts of the substance that he had not been able to collect. Perhaps more important was that he was interested not only in the weights themselves, but in the ratio between the weights of quicklime and slaked lime. Unlike the calcination and reduction experiments, he was not diverted by the example of Stephen Hales to report volume ratios. He could not have been, because there were no volumes measured. Thus the very simplicity of the situation fixed his attention on fundamental premises for a quantitative interpretation of chemical changes.

Lavoisier was now ready to repeat two of the experiments which stood at the heart of Black's interpretation of the role of fixed air in alkalis and alkaline earths: comparison of the action of an acid on calcareous earth with its action on quicklime. The first of five propositions that Black had stated was:

> If we only separate a quantity of air from lime and alkalis, when we render them caustic they will be found to lose part of their weight in the operation, but will saturate the same quantity of acid as before, and the saturation will be performed without effervescence.[20]

Black reported two experiments that he had performed to "inquire into the truth" of this proposition. He "saturated two drams of chalk with diluted spirit of salt [marine acid]. . . . Seven drams and one grain of the acid finished the dissolution, and the chalk lost two scruples and eight grains of air." In the following experiment,

> two drams of chalk were converted into a perfect quick-lime and lost two scruples and twelve grains in the fire. This quick-lime was slaked or reduced to a milky liquor with an ounce of water, and then dissolved in the same manner, and with the same acid, as the two drams of chalk in the preceding experiment. Six drams, two scruples, and fourteen grains of the acid finished the saturation without any sensible effervescence or loss of weight.

His balances were not perfect, but Black could conclude with confidence that "it appears from these experiments, that no air is separated from quick-lime by an acid, and that chalk saturates nearly the same quantity of acid after it is converted to quick-lime as before."[21]

Deciding to perform the second of Black's experiments first, on May 18, Lavoisier placed 6 ounces of spirit of niter [nitrous acid] into a matras which weighed 5 ounces, 1 gros. He "verified accurately" that the weight of the filled bottle equaled the weight of the empty bottle plus that of the acid. Next he poured lime that he had slaked and calcined the day before into a medicinal vial through a strainer placed on the mouth of the vial. The total weight of the vial, strainer, and contents he found to be just 6 ounces. He then poured lime slowly into the matras until it saturated the acid. There was a "light movement of effervescence with a little disengagement of air," which he attributed to a small portion of the lime having become calcareous earth by absorbing fixed air during the day since he had prepared it. Weighing the vial again and subtracting from the previous weight, he determined that the amount of lime employed for the saturation had been 2 ounces, 6 gros, 46 grains. When he weighed the matras again to find the total weight of the materials employed in the operation, the result, 12 ounces, 6 gros, 5 grains, differed by almost one ounce from the weight of the starting materials. Searching for the source of this discrepancy, he decided that in weighing the vial and lime before the operation, "I had deluded myself and that it weighed only five ounces." Apparently he assumed that he had miscounted the weights on his balance to be off by exactly one ounce. After he had corrected for this "probable error," he was able to produce a satisfactory balance sheet (table 1). He noted that the lime used had been a little too much for exact saturation, leaving 8 to 10 grains at the bottom of the liquid and "a little" in the neck. Guessing that the total of this excess was 17 grains, he estimated that the amount necessary to saturate the "6 ounces of my nitrous acid" was actually 1 ounce, 5 gros, 60 grains of

TABLE 1

Calculation			
Weight of the *eau forte* and matras	11 ounce	1 gros	0 grain
Weight of slaked, calcined lime	1	6	5
Total weight of materials employed	12	7	5
Weight after combination	12	6	45
Loss			31 grains

lime. On this basis he calculated that the saturation proportions of acid and lime were "as 13.83:40, that is, very nearly one-third, and exactly as 346 to 1,000." Finding an error in this calculation, he corrected it twice, ending up with the ratio 13.83:48, or "between one-third and one-fourth, and exactly as 288 is to 1,000."[22]

"The small incertitude that this experiment left prompted me," Lavoisier put down, "to begin it again." This time he made no errors in weighing, he saturated the acid "rather quickly," and found "reason to believe that the saturation was exact," because everything remained dissolved. He drew up another balance sheet (table 2).

TABLE 2

Vessel	4	0	13
Acid	6		
Slaked lime	1	5	36
Total materials employed	11	5	49
Weight after combination	11	5	6
Loss			43

The loss, equivalent to "2 drops of liquid," he noted, was "very exactly" the same as in the preceding experiment. "It required in this experiment 13.5 gros of slaked lime to saturate 48 gros of nitrous acid. That is a little more than a quarter, and I believe these proportions to be very exact. If one took the nitrous acid to be 100, it would require a little more than 28."[23]

Quickly moving on to the other half of Black's paired experiments, Lavoisier conducted "the same operation on chalk previously dried very exactly in an iron ladle":

> I employed the same dose of nitrous acid. There was a lively efferves-cence. It was so strong, and the material swelled so much toward the end, that it was necessary [to transfer it into] a large vessel. There was a loss of at most about 1 gros, which leaked out during the operation. I estimated it at no more than 56 grains.

Again he represented his results in a balance sheet (table 3).

TABLE 3

Calculation			
Nitrous acid	6 ounces	0	0
Chalk employed	2	7	44
Total weight of materials	8	7	44
Weight after the combination	7 ounces	6 gros	60 grains
Loss of liquid during the operation by leakage: 56 [grains]			
[Corrected weight after comb.]	7	7	44
Loss during operation	1 ounce exactly		

It seems evident that Lavoisier chose his estimate of the loss by leakage to make the "loss during the operation" come out at exactly one ounce. This final result of his balance table was not a measure of error, but of the quantity of fixed air disengaged during the operation. Like Black, Lavoisier did not demonstrate that the weight of the products equaled that of the starting materials, but assumed that equality, and used it to calculate the weight of the fixed air that he had not measured.[24]

Noticing again that the operation had "sensibly passed the point of saturation [by] around 1 or 2 gros," Lavoisier estimated that the quantity of chalk necessary to saturate the 6 ounces of acid was actually 2 ounces, 6 gros, or 22 gros. "The ratio is thus as 22 to 48, or as 458 is to 1,000. The diminution of weight is thus one ounce, or 8 gros for 22 [gros]—that is, as 364 to 1,000."[25]

Comparing now the respective quantities of chalk and of quicklime that he had found to saturate the same quantity of acid, Lavoisier drew the final conclusion from these experiments: "It follows from this that 22 gros of chalk contains only 14 gros of calcareous earth [he should have written "quicklime"] and water, and the rest is elastic fluid."[26]

Despite the small corrections made by guessing at losses of material and at how far he had missed the saturation mixture, it should be obvious that in this set of experiments on the action of quicklime and calcareous earth on acids, Lavoisier took more trouble to obtain results which were as reliable as he could make them, than at any preceding point in the research program he had begun three months before; that he satisfied himself for the first time that he had achieved a "very exact" result; and that the form of his measurements and reasoning manifest, more fully than anything that had come before it, what has come to be viewed historically as his balance sheet method. It is no coincidence, I believe, that all these features emerged together at a time when he was engaged in the repetition of experiments performed by Joseph Black.

* CHAPTER SEVEN *

Caution and Consolidation

ON MAY 19, the day after performing the experiments modeled on those of Black, Lavoisier was at the Academy of Sciences, continuing to read portions of his history of experiments on fixed air. The next day he was busy again in his laboratory. Although he had modified Black's experiments by employing nitrous air in place of marine acid, and had applied his results to calculate quantitative proportions beyond those reported by Black, he had not moved significantly beyond the methodology Black had applied to the problem. Like his predecessor, Lavoisier had traced the transfers of fixed air only indirectly, through the weight gains and losses of the substances to which Black had assigned the elastic fluid. Now Lavoisier began to search for means to extract fixed air directly from these substances. As in his experiments with volatile alkali and minium in February, he turned first for this purpose to the vacuum pump:

> I placed quicklime and water under the pneumatic machine and produced a vacuum. The lime had been slaked as usual, and reduced with heat to a pulp. I did not notice whether there was a production or absorption of air in this operation, but because the recipient luted to the pneumatic machine allowed air to enter, this experiment should be begun again.[1]

It is not clear what Lavoisier expected from this operation. Perhaps he was testing the possibility he had raised as a result of the surprisingly small loss of weight in his earlier preparation of slaked desiccated lime, that it had absorbed fixed air (see above, p. 53). In any case he did not carry out his intention to repeat the experiment. Instead, he used the pneumatic apparatus to continue the experiments on volatile liquids that he had begun ten days earlier. This time he attained a more interesting, if unanticipated, result. He placed a sample of spirit of wine in a bottle under the recipient of the machine, together with a barometer. As he lowered the pressure, the liquid boiled a little, in puffs. He was able to lower the mercury far enough to make the spirit of wine disappear. Even then, he could not be sure about the effects of the vaporization on the pressure: "It appeared to me that the mercury descended a little during the ebullition, but very little, and it certainly did not ascend." What struck him as "very remarkable," however, was that "there was considerable evaporation during this operation and the spirit of wine collected in drops on the bottle, where it underwent a rather large cooling."[2] Lavoisier did not comment further on this observation, or follow it up. It was, however, the first indication of a phenomenon that eventually he used to support his view that the vapor state is produced by the combination of a liquid with matter of fire.

On this same day, Lavoisier returned to some unresolved problems that had risen during his first efforts in February to repeat Black's experiments on the action of chalk on alkalis. He had saved the earth remaining on the filter after he had mixed lime with soda in an unsuccessful effort to make the alkali caustic (see chap. 3, p. 18). Still impregnated with caustic alkali, it had turned almost to the consistency of chalk, but less hard. "According to the theory" of Black, he had reason to believe that the substance was now "a calcareous earth rather than a lime"; that is, that it should have absorbed the fixed air from the alkali.

> To assure myself of that, I put it in a large quantity of water, but I noticed that it dissolved, and that it was lime. That fact is peculiar, because the caustic alkali obtained from that operation itself gave a light effervescence with acids. This demands clarification.[3]

Still unable to supply an explanation for these irregularities, Lavoisier turned to the further examination of the material left over from a more recent operation. Retrieving the solution of calcareous earth in nitrous acid resulting from the experiment conducted four days before, he divided it into several glasses to try various precipitation experiments. One of them he poured into "phlogisticated alkali" (an ordinary non-caustic alkali), and received "a very beautiful blue precipitate, but the calcareous earth was not precipitated at all."[4] Again matters had not gone for Lavoisier as they should according to Black, who had "shewn . . . that absorbent earths lose their air when they are joined to an acid; but recover it if, separated again from that acid, by means of an ordinary alkali: the air passing from the alkali to the earth."[5]

The rest of Lavoisier's precipitations were less anomalous. The second portion of his solution he put into alkali made caustic with quicklime, and received a slightly yellowish powdery precipitate. The third, put into the alkali of soda, caused a similar, but less powdery precipitate. The fourth portion, placed in an alkali of soda previously boiled with minium gave a precipitate exactly like that given by soda alone. The caustic volatile alkali produced from sal ammoniac by minium made his solution slightly turbid, without producing a precipitate. During the night a thick pellicle of cream of lime formed on the surface. "Finally, ordinary fixed alkali gave [with his solution] a white, slightly yellowish precipitate, almost like soda" had given.[6] In all of these operations, carried out qualitatively, Lavoisier was simply performing variations on the experiments through which Black had shown that the lime left by the action of acids on calcareous earth can recover its fixed air from ordinary alkalis, but not from caustic alkalis, which have none.

Checking on the progress of the fermentation experiments he had begun on May 15, Lavoisier noticed on the evening of the 20th that the first wheat starch vial had caused the water to descend about 9 1/2 lines. The mixture placed in the other bottle had begun to have a strong odor, but "nothing announced acidity, and it did not redden blue paper at all." The situation remained about the same the next day. In the apparatus connected to the bran mixture, in which the water had also descended 9 lines by the 20th, it seemed on the 21st to have reascended slightly, "but almost insensibly." Apparently deciding that nothing decisive would happen in the first two bottles, he dismantled them, but continued to observe what might happen in the bottle containing bran. The red wine placed in the fourth bottle had produced "several lines" of air at first, but by the 21st it had ceased to produce more. Nevertheless, he continued that experiment also.[7]

On the 21st Lavoisier also turned back once again to reexamine the outcome of an earlier experiment. The attempt he had made beginning on April 9 to detonate niter and sulfur under a glass jar had been less than successful (see chap. 3, pp. 28–29). For some reason he was, nevertheless, curious about the properties of the air remaining in the vessel after six weeks. By means of an iron syphon and a pump, he was able to pass the air through lime water, which it caused to become cloudy, "but slowly, much less than [does] the fixed air that leaves calcareous earths during effervescences. Nevertheless, I succeeded in having a light precipitate, and it is probable that in continuing it for a longer time I could have precipitated all of the lime."[8]

When he passed the same air into distilled water, "it seemed to me to impregnate it very little." Neither did it precipitate silver from a solution in nitrous acid, or act on fixed alkali. To test the effect of the air on a burning candle, Lavoisier put together an apparatus with which he could pump the air from the inverted jar in which he had stored it into a tall inverted bottle filled with water, which it displaced (see fig. 7). Putting his hand over the mouth of the bottle after it had filled with the air, he lifted it out, turned it over, and inserted a burning candle suspended from a long handle that enabled him to lower it as far as he wished into the bottle. The candle was extinguished even before entering the bottle. When he inserted it a second time, however, he could lower it two inches before the flame went out. Soon afterward the candle kept burning even when he lowered it all the way to the bottom—"Which proves," he remarked, that "the atmospheric air mixes in a short time with the infected air."[9]

Lavoisier's choice of this particular air to test was probably fortuitous. The idea for subjecting the air produced by a chemical change, or remaining after one had occurred in it, to a series of such tests most likely came from the publications of Joseph Priestley that Lavoisier had read while preparing his history of experiments on elastic airs. Priestley regularly tried the effects of the airs he studied on the burning of candles, on the respiration of small animals, and on lime water, and tested their solubility in ordinary water.[10] The tests that Lavoisier made of the air on solutions of silver and alkali were probably intended to check for the presence of the "acid air," or vaporous marine acid that Priestley had discovered, which would have precipitated the silver, and neutralized the alkali. Lavoisier drew no conclusions from these tests concerning the identity of his air. His belief that he could have precipitated all of the lime water if he had continued longer to pass the air through it reflects his presupposition that it was fixed air that was absorbed and released in all the processes he was studying. That the air also extinguished the candle before mixing with ordinary air would have supported that view, but its inability to impregnate ordinary water was in contradiction with the known properties of fixed air. Up until now, Lavoisier had scarcely considered any other possibilities than that fixed air or atmospheric air itself combined with metals in calcination and inflammable bodies in combustion. His encounter with an air whose combination of properties fit neither species may have given him an early hint that there might be in the atmosphere another elastic fluid of unknown identity.

On the same day that he performed these tests, Lavoisier completed the reading of his history of the elastic fluids at the Academy. He must also have been, for some time, preparing to perform an experiment central to the consolidation of his "new theory"; that is, the reduction

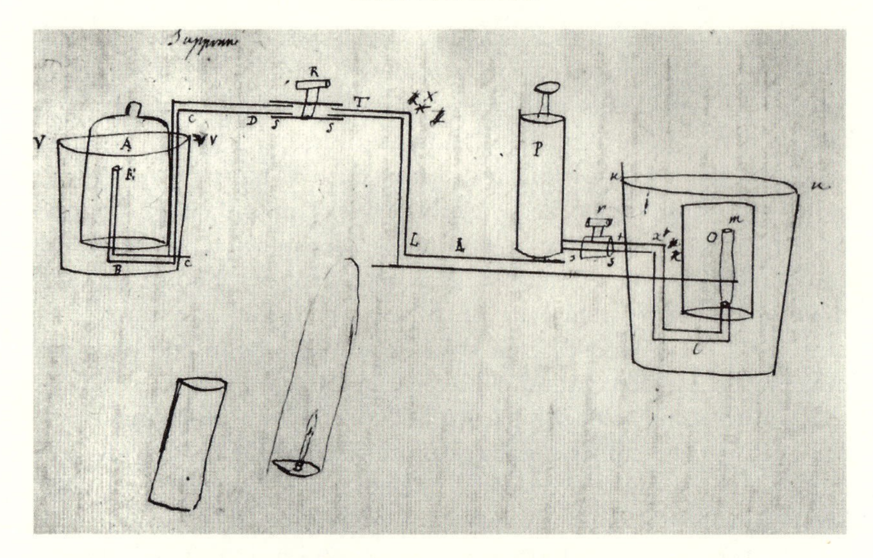

7. Sketch drawn by Lavoisier in his laboratory notebook of apparatus to be constructed. Reproduced from Lavoisier, Cahier lab. 1, facing p. 52.

of lead ore heated in the special retort that was being constructed for him. By this time he had realized that the burning lens arrangement with which he had previously reduced minium was not suitable to obtain accurate results in any of the calcination or reduction experiments for which he had been using it prior to the public meeting at which he had declared that he would later present the details of those experiments. The narrow focus of the burning lens allowed him to operate only on small quantities of metal or calx; but because the heat surrounding the focus was very great, he could not use jars less than 5 to 6 inches in diameter without causing them to crack. The few cubic inches of air disengaged were, therefore, spread out in a much larger space. A very slight change in temperature would be enough to cause "sensible errors."[11] Not only did his recognition of these drawbacks in his method leave him "anxious" about the accuracy of the result he had obtained for the reduction of minium, but must have led him also to see why he had obtained no decisive results in the other experiments he had tried during that hurried time. It must now have become the most urgent matter on his research agenda to repeat these experiments using the much larger quantities of substance he could heat in the rugged iron retort.

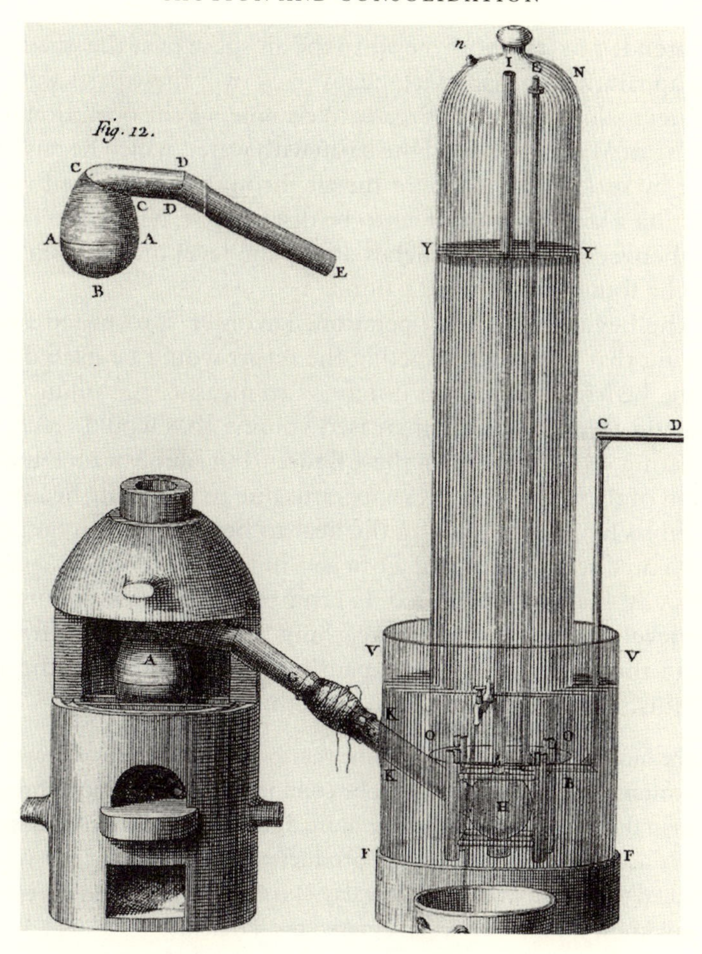

8. Apparatus used by Lavoisier for experiments on reduction of minium. Reproduced from *Opuscules*, Plate II, Figs. 10, 12.

Lavoisier put into his retort 6 ounces of minium—24 times as much as he had been able to use with the burning glass—and 4 gros of charcoal calcined to a powder. Fitting the retort into a reverberatory furnace whose close-fitting dome would allow him to apply heat to all walls of the vessel, he luted the downward slanting neck to a receiver made of white iron. Because it was essential that this connection be air-tight, he used a fatty lute with a firm consistency, covered it with a damp bladder, and wrapped the whole tightly with thread. The receiver rested in a large basin, and was connected in turn to a vertical pipe intended to transport the air to the air space over the water in the Hales apparatus. The jar which he inverted over these parts was about 2 1/2 feet tall. When he was ready to commence the operation, in the morning of May 23, he filled the basin with water, which he raised into the jar by sucking out some of the air through a long metal syphon. Expecting a large volume of air to be disengaged, he tried to raise the water between 28 and 30 inches above the level in the basin, a task which he found difficult and painful.[12]

At the beginning of the operation, Lavoisier "proceeded slowly." Realizing that the air contained in the retort would be dilated by the heating, he left a space in his notebook to measure the volume of the retort and to calculate the increased volume that would result from the heating. He did not fill in these figures, but simply waited until the dilation of the air in the Hales apparatus due to moderate heating had reached its limit before raising the heat to begin the reduction of the calx. "The disengagement of air began to take place" and, starting at 11:06 A.M., Lavoisier recorded its progress every three minutes. The water level descended steadily, reaching 10 inches, 6 lines by noon. Despite the regularity of the expansion of the volume in the Hales apparatus, Lavoisier worried that something was wrong:

> As the lute was beginning to become very hot, anxiety overcame me that the volume of air produced might be coming from the exterior [through leaks in the lute]. I bound the lute more tightly with thread. At the same time the fire diminished and the production of air ceased. The vessels retained the void for the rest of the day. But I always have the uncertainty of not knowing that no air has entered from outside.

His uncertainty did not prevent Lavoisier from assessing the result of the experiment on the assumption that the air produced did derive from the calx. Having already "reduced" quantities measured as inches of descent on the vessel to cubic inches (probably based on the diameter of the vessel), he now made a quick calculation of the ratio of lead formed to air disengaged. "The quantity of lead that should result from this experiment would be about 3/4 of a cubic inch, from which it would follow that in being converted to minium, lead is charged with 388 times its volume of air."[13]

Because he had not yet dismantled the apparatus, Lavoisier's estimate of the quantity of lead must have been a rough calculation derived from the known density of the metal, the difference in weight between lead and its calx determined by Guyton de Morveau, and the weight of the minium he had put into the retort. It is most striking that, even after the experiments on calcareous earth in which, following Joseph Black's example, Lavoisier had analyzed his results in terms of the balance of weights before and after the operation, here he resorted still to a volume comparison like those of Stephen Hales. One practical reason that he did not immediately convert to weights was, undoubtedly, that he did not know the density of the fixed air whose volume he was measuring. A deeper reason may be that he had still not generalized in his mind that weight relations were more fundamental than volume relations; that he was still modeling his approach on that of the respective originators of the various types of chemical operation that he performed.

Recognizing that his calculation was not "scrupulously exact," because he had measured the air produced while it was still warm, and "consequently occupied a larger volume," he waited for the vessel to cool, then measured "exactly the height at which the water stopped" rising. Before disassembling his apparatus, he "wanted to examine the air" produced. To do so, he connected the intake side of "my pump" to the syphon, and the outlet side to a tube connected to a series of three bottles of lime water and an inverted flask. The resulting apparatus was like the third of those he had designed on May 7, with the large Hales apparatus substituted for the small inverted bottle shown there (see fig. 6, p. 51). At the first stroke of the piston, the first bottle began to be clouded. Only after the third or fourth stroke did the second bottle take on the same appearance, the third bottle still later, from which he inferred that not all of the air was fixed air. The excess air passed into the inverted bottle. When it seemed that all of the lime water in the three bottles containing it had been precipitated, he stopped pumping.[14]

Afterward, Lavoisier made a complex calculation, whose purpose is not immediately evident. He had withdrawn from the Hales vessel 4 inches, 5 lines of air, which corresponded to 120 cubic inches. By weighing, both before and after the operation, the inverted bottle into which the air not absorbed by the lime passed, he found that the air collected there weighed an amount equivalent to 82 cubic inches of ordinary air. From the difference between that volume and the volume removed from the other vessel, he inferred that "there was, therefore, a loss of 38 cubic inches, which was combined with the lime. Supposing that the density of fixed air is the same as that of the atmosphere, that corresponds to a weight of 18 to 20 grains."[15]

Here Lavoisier seemed to be feeling his way toward a means to convert his volume relations to weight relations. His calculation rested on the unconfirmed assumption that the density of fixed air was not significantly different from atmospheric air. It is remarkable that he did not try to determine the weight of the fixed air absorbed in his lime water bottles by the much more direct route of weighing the bottles before and after the operation. That he did not do so reinforces the impression that he did not have, in advance, a clear conception that the weight relations were primary. To calculate the weight of the fixed air in this roundabout manner, when it was too late to measure it directly, suggests that, for Lavoisier, this determination was only an afterthought.

After taking his apparatus apart, Lavoisier turned the Hales vessel right-side up, poured water into it up to the mark he had made of the initial level, emptied the water into another container, and poured water in again to the level of the next mark he had made. Repeating this process for each of the marks representing the level at the successive stages of the operation, he weighed each portion of water. From the known density of the water he could then calculate the volume of the air at these stages. To make such calculations more convenient, he wrote, on the normally blank left-side of the laboratory notebook, a "Table of the volumes of air corresponding to a certain weight of water."[16]

"I wanted to put more exactitude into the results of my operation of May 23," Lavoisier wrote on the day after. To that end he measured the volumes of the retort, the receiver, and the vertical tube, and the dilation produced by the initial heating, all by filling the spaces with water and then weighing the water. The total of the three spaces in the apparatus was 94.6 cubic inches, and the expansion due to the heat was 31 2/3 cubic inches. The weight of the water necessary to fill the space in the inverted jar between the level of the water at the commencement of the operation and the level at the end after cooling was equivalent to 256 1/4 cubic inches. "That is," he wrote, "the quantity of air produced." Thinking that he could find the amount of air produced in another way, from the space between the mark made after the initial heating and the mark made at the end of the operation but before the vessel had cooled, he calculated on that basis a volume of 294 cubic inches. That, he inferred, "is too much, because one is starting from dilated air." After writing down a summary of these calculations to evaluate the volume of the air produced, he added casually at the end of his account of the experiment that "the quantity of air disengaged is around 2 gros, more or less."[17]

It is hard not to be struck by the disparity between the exertions Lavoisier made to determine as accurately as he could the volume of air disengaged in the reduction of the minium in this experiment, and the absence of a clear reason for the achievement of that accuracy. Not only did he treat the weight of the air in question as secondary, but he did not balance his attention to the quantity of air released with corresponding attention to the quantity of material lost in the reduction of the calx to the metal. So far as can be told from his record of the experiment, he did not collect or weigh the lead obtained in the operation. It is possible that he merely postponed that task, retaining the material for later examination, as he had done in some of his earlier experiments. The overall impression, however, is that, having finally attained a method capable of measuring accurately the air released in the reduction of a metallic calx, he did not know quite what to do with the result.

Instead of pursuing the problems raised by this experiment, Lavoisier again changed direction, returning to the phenomena related to chalk, alkalis, and the exchange of fixed air between them, and taking up experimentally for the first time a phenomenon involving fulminating gold that he had discussed in his Easter memoir. That he placed higher priority, just at this point, on the examination of these phenomena than on perfecting the experiment most central to his own theoretical structure, is surprising enough to invite some surmises about his motivations and his state of mind. The most plausible explanation is that he was not only uncertain about the treatment to give to his measurements of the reduction of minium, but puzzled about the nature of the air released. He had measured two quantities: that of the fixed air absorbed in lime water, and that of the total disengaged from the minium. The former was only a small portion of the latter. What was the rest of it? According to the theory he had announced so confidently in April, it should all be fixed air. Unable to reconcile his theory with his observation, he may simply have decided to put the question aside for awhile. It might have seemed safer, in the meantime, to move on to phenomena for which there seemed less doubt that the elastic fluid exchanged really was fixed air.

"Fulminating gold," P. J. Macquer had described in his well-known *Dictionnaire de Chymie* in 1766 as a "preparation which is a precipitate of gold from its solution in *eau régale* [a mixture of marine acid and nitrous acid]: it has been named *fulminating* because, when it is heated, or merely rubbed to a certain degree, it gives an explosion comparable to, and perhaps even superior to, that of [detonating] powder itself." The phenomenon was, he wrote, "one of the most striking and most marvelous that chemistry offers us." Later in his article he added that it was also one of the most violent and dangerous phenomena, one that had killed or injured unwary chemists. The cause of the fulmination was not easy to discover, according to Macquer, but from the facts that a solution of gold in *eau régale* did not fulminate unless sal ammoniac was present or volatile alkali had been used in its preparation, and that the weight of the precipitate exceeded by a quarter that of the gold in solution, he inferred that a nitrous ammonia combined with the gold. The detonation was due to the "great quantity of inflammable matter that the volatile alkali contains." Antoine Baumé had given a different explanation, based on the formation of a "nitrous sulfur" by the union of phlogiston with nitrous acid.[18] Lavoisier had adduced fulminating gold in his Easter memoir to illustrate his contention that "all metallic substances, without exception, are susceptible to combine with fixed air." In this case, he claimed, "a heat only a little above that of boiling water is sufficient to disengage [from a metallic calx] its fixed air and cause the reduction." The formation of the precipitate, he pointed out, required a "body, such as sal ammoniac, or volatile alkali, which contains fixed air in great abundance." Precipitates prepared without those substances do not fulminate.[19] The phenomena that Macquer had ascribed to the inflammable matter in sal ammoniac and volatile alkali Lavoisier thus attributed to the fixed air that they contain. He had probably not based this claim on experiments he had himself performed, but had inferred the role of fixed air in the process by connecting the known operations for the production of fulminating gold with Black's experiments on volatile alkali. Now he decided to examine the process more closely.

Apparently Lavoisier did not feel the need to verify the formation of fulminating gold through a process that supplied fixed air from volatile alkali, but planned immediately a counterproof for his interpretation: that is, a precipitate of the gold made by adding caustic volatile alkali to a solution of gold in *aqua regia* should not fulminate, because the caustic alkali should not contain fixed air. To begin, he prepared volatile caustic alkali by the action of slaked lime on sal ammoniac. Mixing them with water in a clay retort, he distilled off the alkali and collected it in liquid form. At the same time he made his solution of gold in aqua regia. He weighed the quantities of all the substances involved. "I took one ounce of the solution," he wrote, "and I precipitated it with my caustic volatile alkali." The process went quietly, producing a pale yellow precipitate. At the end it became a little darker, an effect he attributed to an excess of volatile alkali which had "no doubt, furnished air." Placed on a knife blade, the precipitate "detonated a little, like gunpowder, but without fulmination." Heating it, he evaporated the nitrous acid quietly, leaving gold in the vessel. "I left this gold drying on the filter," he related, "and I was quite amazed to see afterward that it fulminated as well as any other fulminating gold."[20]

This little explosion shattered one more of Lavoisier's expectations. There is no record of the conclusion he drew, but he must have at least come to doubt the explanation for fulminating gold that he had recently presented. Alternatively, he might have inferred that he had failed to remove all of the fixed air from the sal ammoniac he had used to produce volatile caustic alkali. His next actions suggest that the result may even have prompted him to question whether Joseph Black's theory of causticity was as firmly established as he had supposed while reproducing Black's experiments.

In his historical essay Lavoisier had reviewed not only Black's study of fixed air, but an opposing view originated by a German apothecary named Johann Friedrich Meyer. Lavoisier gave Meyer a very respectable hearing. "While the doctrine of fixed air was establishing itself peacefully in England, he wrote, a redoubtable contradictor arose in Germany." Meyer had proposed, in 1764, that a hitherto unknown acid, which he named *acidum pingue*, was the cause of the causticity of caustic alkalis. Where Black ascribed causticity to the removal of fixed air, Meyer attributed it to the addition of *acidum pingue*. Meyer believed that the same substance entered the calx of metals. That feature was, Lavoisier wrote in his review, where "the system of Meyer seems to have the advantage over the English system." That is, it accounted naturally for the gain in weight in calcination, as well as several other phenomena. "There are few books in chemistry which reveal more genius than that of M. Meyer," Lavoisier wrote. If his ideas were accepted, they would result in a "new theory directly contrary to that of Stahl and of all modern chemists." At the end of his summary of Meyer's views, he did add a cautionary note: "One must acknowledge that this chemist has abandoned himself somewhat to the propensity of all those who believe they have discovered a new agent, and who apply it indiscriminately to everything."[21]

The generous terms in which Lavoisier discussed Meyer's views have helped give Meyer a prominent place in historical accounts of Lavoisier's early career. Guerlac's opinion that Meyer had given an "elaborate and apparently convincing refutation" of Black's work[22] was probably a reflection, in part, of Lavoisier's laudatory language. J. B. Gough went further. Having found some notes written by Lavoisier in 1766, mentioning an experiment that "favors marvelously the system of *acidum pingue*," Gough concluded in 1968 that "Lavoisier was sufficiently impressed by Meyer's ideas to have become an adherent to his chemical system."[23] In 1930, Andrew Meldrum had been more skeptical about Lavoisier's attitude toward Meyer. In his historical survey, according to Meldrum,

> Lavoisier gave a fair account of Black's doctrine and of Meyer's and in doing so he wrote more cordially about Meyer than about Black. In the Experimental Part he almost ignored Meyer: he concurred with Black. One feels that Black's clear ideas made an irresistible appeal to Lavoisier as his experiments proceeded, [and he] found that fixed air was a real thing that he could liberate . . . and transfer.[24]

These somewhat divergent views are not irreconcilable. When Lavoisier first encountered the French translation of Meyer's treatise, which appeared in 1766, at a time when he was not yet familiar with Black's work except through Meyer's opposition to it, he could well have been caught up in youthful enthusiasm for what he still viewed seven years later as a work of genius. By 1773 he had gained enough experience to view the ongoing debate between the "English system" and that of the followers of Meyer from both sides. In his discussions of the works of the authors who had published experiments and arguments in support of Meyer and of Black, respectively, Lavoisier seems to have made a special effort to report the positive aspects of all contributions to the question, and to avoid appearing more favorable to one camp than to the other. It was, in fact, in his interest to present the debate as an open one that called for his own experiments to resolve the issues in question. Meldrum's opinion that in his experiments Lavoisier followed Black and ignored Meyer was based on the published version of the experiments. From the laboratory record we can see that it is only partly right.

Psychologically, Lavoisier might have identified somewhat more with Meyer than he did logically. He, too, was developing a new "system" which he thought might eventually put him in opposition to

Stahl and "all modern chemists." His final comment about Meyer's propensity to apply his new "agent" to "everything" can be viewed also as a subtle warning to himself to be more careful than that.

About a week after his indecisive experiment on fulminating gold, Lavoisier thought of a way "to decide all at one stroke, and in a manner that would leave nothing uncertain, whether lime gives something to a caustic alkali in the fashion of Meyer, or whether, on the contrary, it removes something according to the English hypothesis."[25] His new method of choice relied on an instrument of which he had been enamored five years earlier.

In 1768, after several years spent accompanying Jean-Etienne Guettard on geological expeditions, Lavoisier had viewed himself as a field naturalist, embracing a "vast field . . . of research extending, as far as it was possible, into all parts of the sciences, arts, and natural history."[26] As an ambitious young aspirant to a position in the Academy of Sciences, however, he needed to identify some more specific line of investigation in which he could excel. While his family and other connections sought an institutional opening for him in that august assembly, he believed he had found his scientific opening in the analysis of mineral waters. Such analyses were of interest, not only to "a small languishing portion of society," he intoned in a memoir on the subject intended, among other things, to help gain him admission, but was "of common interest to the entire society." It was also a demanding task. Citing writings by such eminent chemists as Macquer and Andreas Marggraf, and the influential Academician Jean-Baptiste Leroy, as evidence for the inadequacies of quantitative analyses of waters based on precipitations and evaporations, Lavoisier portrayed the current state of the subject as a disaster. Fortunately, he could propose a remedy. "If salts augment the weight of water, if that increase is always in a certain proportion to the quantity of salt added to the water, does that not provide a sure means to know its quantity?" All that was necessary was a "method to determine the density of a water with scrupulous exactitude," and tables expressing for each salt the relation between the density of the solution and the quantity of salt in it. For the former purpose, Lavoisier redesigned an instrument known as the areometer. Those in use before him consisted of a weighted bulb to which was attached a vertical, graduated stem. By noting the mark on the stem to which the instrument sank in fluids of different densities, one could determine the relative densities of the fluids. Lavoisier reversed this relationship. His areometer contained only a single mark on the stem, with a platform on top on which he could place weights. The difference between the weights necessary to cause the instrument to sink to the same mark on the stem in fluids of different density was a direct measure of the differences in weight of the same volume of the fluids (see fig. 9). The main advantage of his method, he claimed, was that it allowed for extreme accuracy without requiring the very long neck that would be necessary to achieve comparable results with a graduated stem instrument.[27]

Lavoisier continued to pursue his interest in hydrometry, the densities of fluids, and the analysis of waters, until his discoveries concerning the fixation and release of airs in the fall of 1772 opened a more auspicious avenue for his scientific ambitions. At the end of May 1773, it appeared to him that his older interest could now come to the aid of his newer venture. The idea was to measure the density of a solution of an ordinary alkali in water, then to add slaked lime to the solution to convert the alkali to caustic alkali. Because the lime would then precipitate out as calcareous earth, the solution at the end would contain only the same amount of caustic alkali that had formerly been ordinary alkali. If, when he measured the density again it had increased, that would indicate that something like Meyer's *acidum pingue* had been transferred from the lime to the alkali. If the density had diminished, that would support Black's view that fixed air had been removed from the alkali, leaving a caustic alkali that weighed less.

After dissolving 2 ounces of crystals of soda in 14 1/2 ounces of water, on June 4, Lavoisier "plunged" a silver areometer weighing 9 ounces, 64 grains into the fluid. To cause it to enter down to the mark on its stem, he had to charge the instrument with 3 gros, 64 1/2 grains of additional weight. He then added 1 ounce of lime that had previously been slaked, dried, powdered, and passed through a silk sieve. After agitating the solution, he let it stand. Within a few minutes the lime had collected at the bottom. He decanted the liquid and dipped his areometer into it again. This time it was necessary to add only 3 gros, 48 1/2 grains of weights to reach the mark. He repeated the operation two more times, the last measurement of the density requiring 3 gros, 21 grains. Afterward he found that the liquid did not effervesce with nitrous acid, whereas the precipitated lime made a "very lively effervescence with acids." Even though he suspected that the lime had not been totally converted into calcareous earth, these tests verified that the alkali had been converted, all or in part, to caustic alkali. Sometime later he wrote across the top of his laboratory record

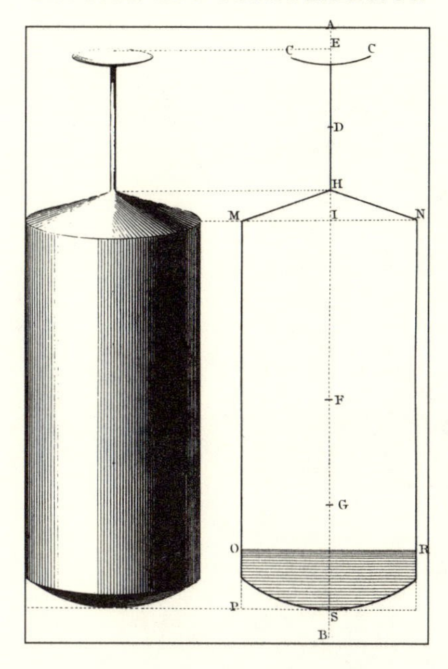

9. Lavoisier's areometer. *Left*, exterior; *right*, cross section. Reproduced from Lavoisier, *Oeuvres*, vol. 3, Plate.

"Diminution of the specific gravity occasioned by lime in an alkaline lye."[28] In terms of the criteria he had set for himself, he had decided the issue in favor of the "English hypothesis."

Lavoisier now returned to the two fundamental experiments of Joseph Black, the comparison of the action of lime and calcareous earth on an acid, that he had already reproduced with detailed balance sheets, on May 18. He was ready this time, however, to go beyond Black's indirect estimates of the fixed air disengaged from the differences in the weights of the other substances involved. By an ingenious device, he had adapted the "great glass vessel" that he had previously used for the reduction of minium with the iron retort, to enable him to perform these operations within the air space at the top of the jar while it was inverted in the water bath. A pedestal rising from a copper base at the bottom of the water basin into the space above the water supported a circular platform on which he could place a jar containing the lime or the chalk. At one edge of the platform was mounted a rectangular frame with two pivots between which a small pitcher could be clamped, so that it could rotate around the horizontal axis between them. The pitcher was weighted so that it hung right-side up with its

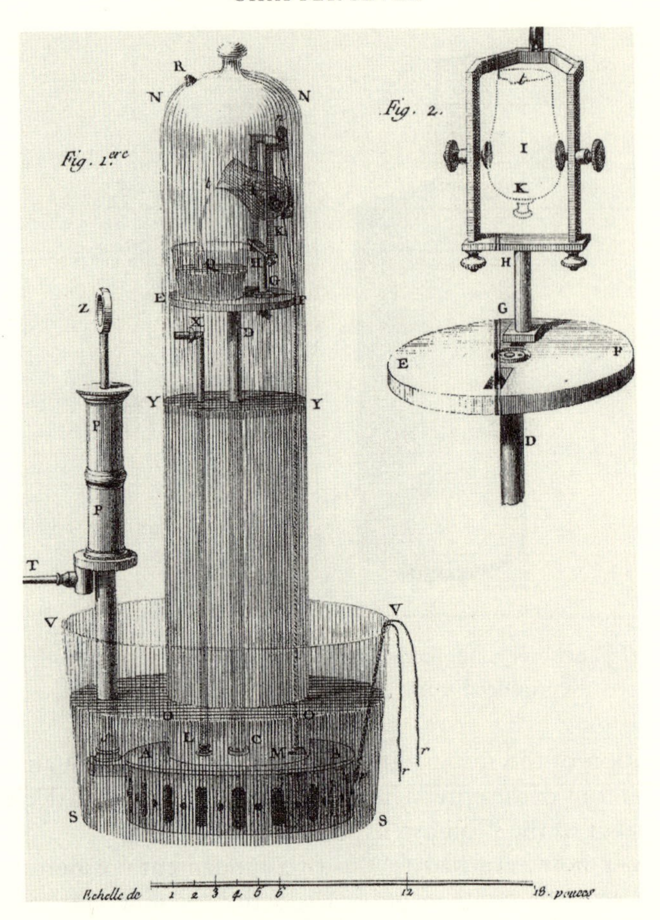

10. Apparatus used by Lavoisier for experiments on action of acids on lime and calcareous earth. Reproduced from *Opuscules*, Plate I, Figs. 1, 6.

contents of nitrous acid. Lavoisier had attached to it, however, a cord and pulley arrangement that enabled him, by pulling on the end of the cord that emerged from the base at the bottom of the inverted vessel, to tilt the pitcher from the outside, so that it would spill its contents into the jar containing the lime or the chalk[29] (see fig. 10). Awkward though this contrivance may appear, it apparently worked. With it, Lavoisier could hope to measure, for the first time, all of the substances supposed by Black to participate in these processes.

On June 7, Lavoisier put 3 gros, 27 grains of the same desiccated slaked lime that he had prepared on May 17 into the jar, and poured 11 gros of nitrous acid, the quantity necessary to saturate the lime, into the pitcher. A small amount of the acid spilled accidentally onto and "attacked" the copper plate of the support. When he had placed the large inverted vessel over the support apparatus, raised the water level, and tipped the acid into the lime, a lively effervescence ensued, producing heat and smoke. He noted, however, that, "although tumultuous, the effervescence was very different from that which takes place with ordinary calcareous earth." The bubbles were very small. The water suddenly descended 3 or 4 inches, but within less than a minute it began to rise again to within 10 or 12 lines of the original mark. An hour later, when the vessel had cooled, the difference was only 5 lines, and by evening only 3 lines. The quantity of air released was, therefore, 7 cubic inches, or 3 1/2 grains. Lavoisier immediately dismantled the apparatus to prepare for the experiment with calcareous earth.[30]

For that purpose, Lavoisier placed in the jar 5 gros, 27 grains of chalk, an amount which he had again determined in advance was necessary to saturate the 1 1/2 ounces of nitrous acid he put in the pitcher. He set up the apparatus as before, but when he began to pour the acid into the chalk, "some circumstances disturbed the apparatus. The mixture was made more rapidly than I wanted, and I was not able to pour in the last portions." Undoubtedly the trouble arose from some kink in the cumbersome cord and pulley arrangement on which he relied to tip the pitcher. About a quarter of the chalk remained, at the end of the flawed operation, undissolved. "Nevertheless, I obtained about 5 inches, 9 lines of air, which were reduced during the night to a little less than 5 inches."[31]

Thinking that he could "correct" this result by taking into account the portions of unused acid and chalk, Lavoisier estimated that "if the experiment had succeeded, I would have had 6 1/2 inches of air, that is, around 180 cubic inches, or 1 gros, 18 grains."[32] To convert the volume to a weight, Lavoisier had assumed the density of fixed air to be equal to that of ordinary air. He was so certain that the air produced was fixed air that he did not pass it through lime water, as he had done with the air produced in the reduction of minium.

Also unlike his treatment of the result of the minium experiment, Lavoisier regarded the measured volume in this experiment only as the means to attain a weight. That was because, having previously constructed a balance sheet of the operation based on the weights of the chalk, the acid, and the liquid remaining after the operation, from which he had *calculated* the weight of the fixed air disengaged, he wished now to compare the measured quantity of fixed air with the quantity he had then reached from the difference between the substances used and the other substance produced. Because he had employed 6 ounces of nitrous acid in the earlier experiment, and 1 1/2 ounces in this one, he multiplied by four the weight of the air that he estimated would have been produced if all the acid had been used, to give the amount that would have been produced in the earlier experiment:

> By quadrupling that quantity [1 gros, 18 grains] one would have 5 gros for the weight of the air disengaged during the operation of 18 May. The loss of weight had been 8 gros, which proves that a notable quantity of fluid had evaporated, or that fixed air is heavier than ordinary air.[33]

The "loss" of 8 gros to which Lavoisier referred was the 1 ounce at which he had arrived at the bottom line of his balance sheet for the experiment of May 18 as the weight of the fixed air produced. The discrepancy between that result and the 5 gros he calculated from the present experiment was what led him to surmise either that some of the fixed air had been lost by evaporation, or that the assumption he had made that the weight of fixed air was the same as ordinary air had given him too small a figure when he converted the volume to a weight.

The possibility that evaporation was the cause of the discrepancy Lavoisier thought he could check by comparing the result with that of the comparable experiment with slaked lime. Correcting for the fact that some of the air produced in that operation might have been reabsorbed, he estimated the weight produced at 5 grains. Quadrupling that figure gave 20 grains, "from which it follows that evaporation of the fluid, despite the great heat can only have been about 20 grains."[34] Although his reasoning is not clear to me, his conclusion seems to have been that evaporation could not explain the difference between his two results for the disengagement of fixed air in the experiments with chalk. Elimination of that possibility pointed to a significant difference between the density of fixed air and ordinary air. Lavoisier probably resisted that inference. As we have seen, he had designed apparatus in March intended to weigh fixed air. Apparently that apparatus still did not exist. To abandon his assumption that he could convert volumes of fixed air to weights by assuming a density approximately equal to that of the atmosphere would mean, therefore, to abandon for the present the attempt to complete the analysis of his experiments.

Viewed from a later perspective, Lavoisier's interpretation of his results appears to have been built on so many assumptions and guesses at the magnitude of errors due to experimental flaws, that agreement between the results of the two experiments would have been fortuitous. Nevertheless, he had, in principle, if not in practice, made another significant methodological advance. For the first time he had incorporated measurements of the quantities of liquids, solids, and air into an effort to arrive at a complete balance between the weights of the material before and after a chemical operation. That his sheet did not balance was less important than that he had tried to make it balance, and that when he failed, he assumed the discrepancy was due to some complicating factors that he had not incorporated into his calculations. Moreover, he was not satisfied with the experiment, and probably began plans immediately to repeat it.

On June 8, perhaps before the preceding experiment was completed, Lavoisier explored qualitatively another set of experiments described initially by Black, in which the action of lime on an alkali could be reversed. "If caustic alkali be destitute of air," Black had reported, it will "separate a calcareous earth from acids under the form of a calcareous earth destitute of air, but saturated with water, or under the form of slaked lime."[35] This phenomenon had been studied more extensively by Nicolas Jacquin of Vienna, in an examination of the theories of Black and Meyer that Lavoisier had reviewed with special admiration in one of the historical memoirs he had read at the Academy. In his summary of Jacquin's experiments on this subject, Lavoisier had described them as a "means to make lime by the wet way."[36] He now tried this operation for himself. He dissolved chalk in nitrous acid, diluted it with distilled water, and "precipitated it with a caustic fixed alkali which [had] made only an almost insensible effervescence with acids." When he tried the same operation with volatile caustic acid, he did not obtain a precipitate, making him think that there may not have been any calcareous earth dissolved in the nitrous acid. He eliminated that explanation, however, by precipitating the substance with an ordinary fixed alkali and finding that during the night a pellicle of slaked lime had formed.[37] He did not, for now, explore this more "complicated combination" further, returning instead to his quantitative investigation of the combinations of chalk and lime with nitrous acid.

Having had a glass "pese liqueur" of the type designed by Fahrenheit made for him, Lavoisier measured the specific gravity of nitrous acid and of limewater. On June 19 he repeated, with this instrument, the measurement of the diminution of the density of fixed alkali when it is made caustic by combination with desiccated, slaked lime that he had already performed to settle the debate between the system of Meyer and that of Black.[38]

The more demanding business for that day was to repeat the measurement of the air released by the action of nitrous acid on calcareous earth in his large pneumatic vessel. Before beginning the experiment he had added a new component to his apparatus to overcome the difficulty he had experienced raising the water into the inverted jar by sucking the air out through a syphon. He had a copper base constructed, on which the jar rested in the water basin. A vertical tube attached to this base, with its open end in the air space just below the platform, was connected at its lower end to the inlet end of Lavoisier's air pump. With this attachment he could easily elevate the level of the water up to 30 inches above the level in the basin.[39]

Lavoisier used the same quantities of chalk and nitrous acid as in his first attempt—5 gros, 27 grains of the former, and 1 1/2 ounces of the latter. After elevating the water with the pump, he covered it with a layer of oil and marked the level of the surface. This time his cord and pulleys worked smoothly. "I made the combination little by little," he recorded. The water descended 7 1/2 inches during the first quarter hour, and only 3 lines further before it came to rest. He allowed the apparatus to stand for two days, while the level remained constant. The corresponding volume was 201 cubic inches. Noticing that there was a little chalk left undissolved, he corrected the quantity necessary for saturation to "5 gros, 12 grains at most." Otherwise there seemed this time to be no operational errors requiring him to adjust his measured result for the fixed air released.[40]

Adjustments were still necessary, however, to fit this result to the experiment in which he had weighed the rest of the substances involved. Two hundred cubic inches still fell short of the 1 ounce (or 8 gros) loss of weight shown in the original balance. Probably because he had more confidence in the second experiment carried out under his inverted vessel than in the first one, he now adjusted the original experiment to this one, rather than the other way around. Dividing the weight loss determined there by four, to make it correspond to the amounts used in the new experiment, he persuaded himself to "believe that" part of the weight loss was due to "the dissipation of water":

> Supposing that it accounts for 1/2 gros of the diminution of weight, there remains 1 1/2 gros for the weight of the air disengaged during the present operation. Now 1 1/2 gros of air in the summer corresponds to a little less than 200 cubic inches, from which it follows that fixed air is almost equiponderable with the common air that we breathe.[41]

Lavoisier seemed quite satisfied with this outcome. He had been able to produce what appeared to be a complete balance of the materials before and after the operation. The result was a quantitative verification of the experiment most central to Black's theory of fixed air. Moreover, he had achieved that end without having to accept the troublesome complication that the density of fixed air might be different from ordinary air. That he had done so by choosing arbitrarily to attribute to a loss of water that he had not detected experimentally, just the quantity needed to make his balance come out, did not seem to concern him.

While Lavoisier worked relentlessly through the summer of 1773 to produce the experimental evidence for his new theory that he had promised in his announcement of the theory in April to reveal in subsequent meetings of the Academy, word of his claims must have been spreading through the European scientific community. Historians have uncovered remarkably little information about the initial response to his announcement, but there is, in a letter from Benjamin Franklin to the Academician Jean-Baptiste Leroy, a clue that it was both widespread and skeptical. "We have nothing new here in the philosophic way," Franklin wrote on June 22. "I should like to hear how M. Lavoisier's doctrine supports itself, as I suppose it will be controverted."[42]

✻ CHAPTER EIGHT ✻

The Long Summer Campaign

ON JUNE 27, Lavoisier went back, for the first time since he had deposited the sealed note in November 1772, to the experiment that had first inspired his program to study the fixation and release of airs: the combustion of phosphorus. Although he appears to have regarded the phenomenon of weight gain that he had observed then as the most secure foundation of his theory, he probably wanted now to determine the proportions of phosphorus and air combined in the resulting phosphorous acid. (The experiment in a vessel saturated with water vapor that had convinced him that the gain derived from the fixation of air rather than the absorption of water had given him only a minimum value; see chap. 2, p. 12).

In his first attempt, Lavoisier resorted to the burning glass and small jar inverted over a water basin that had yielded such meager results for him during the spring. In this case, he had only to use the sun's rays to ignite the phosphorus, but his method nevertheless constrained him to employ only 8 grains of the material. Perhaps to reduce the ratio between the total air space and the air he expected to be absorbed, he used a jar of only 4 1/3 inches diameter. As before, he covered the water with a layer of oil. When the phosphorus began to burn, the water level descended instantly, dilated by the heat produced, but it soon began to rise again. When the combustion was finished and the vessel had cooled, it stood at 1 inch, 6 lines above the original level, "which gives an absorption of 22 1/3 cubic inches." A small amount of yellow material was left over. Lavoisier estimated it at "2 grains at most," which he would have to subtract from the 8 grains to give the quantity burned. He noted also that there had been "much humidity around." The acid produced formed whitish vapors that collected on the interior surface of the jar and attracted the humidity until it formed, within a few minutes, drops of clear liquid. It was this deliquescent property of phosphoric acid that had made it so difficult for Lavoisier to decide, in his initial experiments of the preceding fall, how much of the weight gain to ascribe to air, and how much to water. Here too, it prevented a direct measurement, but he could at least determine the ratio of the air absorbed to the phosphorus burned. In the published version of his description of the experiment he left that result in the mixed form of "around three cubic inches for each grain of phosphorus," a hint that he considered the experiment too inaccurate to merit converting the result to a uniform measure.[1]

Lavoisier intended to continue to improve the experiment on the combustion of phosphorus, but he interrupted that effort the next day to repeat, for Trudaine, the reduction of lead calx with the iron retort and large pneumatic collecting jar. Trudaine may well have urged upon him the necessity to return to an experiment whose first result had been ambiguous at best. Lavoisier later credited Trudaine with having "more than once guided me in the choice of experiments," having "often clarified for me their consequences," and having "wanted to have most of them repeated under your eyes."[2] If the tribute was genuine, as there is reason to believe it was, then this experiment may be one we can identify with the guidance Trudaine gave the younger man. It is easy to imagine that he might have advised Lavoisier that he could not leave the situation in the unfinished state that had resulted from the original experiment.

The second attempt took place on June 28. Lavoisier changed the proportions of minium and charcoal, mixing this time 6 gros of the latter, instead of 4 gros, with the same amount, 6 ounces, of minium. This was a larger proportion than necessary to reduce the ore. His reason for adding the excess was that in the previous experiment, the lead had been found afterward in a single mass difficult to remove from the retort. That may, in fact, have been one reason that Lavoisier had not weighed the solid product of that operation. He hoped now that the lead might form in grains surrounded by undecomposed charcoal.[3] The reduction began well enough, as the "disengagement of air took place in the ordinary way." But trouble soon intervened: "Having given a blast of the fire a little more than ordinary toward the end," he wrote, "the disengagement became so abundant, that I believed the solder in the iron retort had melted." He stopped the operation, examined the retort, and saw that "I had been alarmed without cause." Unfortunately he also observed, from the appearance of the material in the retort, "that I did not reach my goal [je n'ay pas ete jusqu'au bout]." Because he also had too little time to allow the apparatus to cool sufficiently, he had "uncertainty about the quantity of air disengaged, but I had reason to believe that it was between 15 and 16 inches measured along the height of my vessel, which would give 415 to 440 cubic inches, and at least 2 1/2 gros in weight, supposing the air equiponderable with common air."[4]

Despite another defective operation, Lavoisier was much clearer than in the first experiment about what the aim of his measurement was. Weighing the material left in the retort afterward, he found "only 6 ounces, 1 gros, 22 grains." Adding to that the 2 1/2 gros he had calculated for the weight of the air, he obtained for the total of the materials after the operation, 6 ounces, 3 gros, 58 grains. "There had been, consequently, 2 gros, 14 grains of loss." Given the experimental uncertainties, that was small. Lavoisier must have sensed himself to be close to achieving a satisfactory quantitative account of the reduction on which the case for his theory rested so heavily. He made, however, one observation that probably suggested to him that he had not accounted for all of the changes that had taken place:

> In the ball of white iron that came out of the receiver there were found several drops of water that were neither acid nor alkaline. A little bit was also in the tube mounted on the receiver, and some vapors had condensed on the walls of the receiver.[5]

Either then, or sometime not long afterward, it occurred to him that this water might have been separated from the minium, and that it might even account for the weight loss that he had found.[6]

Everything seemed to point to the desirability of repeating this critical experiment again. Lavoisier had come close enough to balancing the materials before and after the operation to suggest that a fully satisfactory balance was within reach. He turned instead back to his phosphorus experiments. It is hard to escape the impression that the method he had devised for the reduction of metallic calxes was tedious enough to prepare and carry out that he was not eager to go through the whole process again, before getting on with some of the easier experiments on his agenda.

When he returned to his phosphorus experiment, further surprises awaited him. Having allowed the jar in which he had carried out the combustion to stand for two days, and finding the volume of air unchanged, he became "curious to examine this air." Passing it into a tall bottle in the same manner that he had handled the air left from the detonation of niter and sulfur five weeks earlier (see chap. 7, pp. 62–63), he lowered a candle into the bottle several times, as he had done then. Each time the candle was extinguished, but it was necessary each time to go deeper. It seemed to him that "this air mixed a little more promptly than fixed air with the air of the atmosphere." Then he "tossed" a bird into a bottle of the same air:

> It rested there for several seconds without seeming to suffer. . . . It did not appear to have difficulty with its respiration, nor a commencement of convulsions. If it had remained for the same length of time in fixed air, it would have been dead on the spot.

The decisive test of whether the air was, or was not fixed air, was to bubble it through lime water. Lavoisier tried it. No precipitate appeared. He tasted the lime water attentively, and thought that it might have been slightly softened.[7]

By this time Lavoisier must have been thoroughly puzzled. Again, and more strikingly than before, he had encountered an air that seemed to be, in one or another of its properties, unlike both fixed air and ordinary air. What else could it be? Knowing of no other kind of air with which he could identify it, Lavoisier had the idea that it might be ordinary air deprived of the fixed air that was ordinarily present in it: "Persuaded that the combustion of phosphorus absorbs the fixed air contained in the air, or, rather, suspecting it, I thought that by rendering fixed air to this air [in which he had performed the combustion] I could restore the common air." He mixed with the air remaining, one-third its volume of fixed air, but "that mixture extinguished a small candle even more promptly, it appeared to me, than the air in which one has burned phosphorus alone." He concluded that he had "introduced too much fixed air,"[8] but he did not repeat the mixture with less fixed air. Maybe he had used up the air left from the combustion.

Lavoisier decided to repeat the combustion of phosphorus, but in a jar inverted over mercury. From his study of Joseph Priestley's publications he had learned that Priestley used mercury in his pneumatic troughs for the same purpose that Lavoisier had resorted to a layer of oil over water—to keep fixed air from dissolving in the water. Whether he had wanted for some time to adopt Priestley's method, or whether he employed it in this case specifically because he wanted to avoid humidity in burning phosphorus, he did not state. The main obstacle to overcome may have been to procure enough mercury: for this experiment he required twenty pounds of it. He used the same small quantity of phosphorus, 8 grains, as in the preceding experiment, and ignited it in the same way with the burning glass. The phosphorus seemed to him to burn a little better over the mercury than it had over the oil and water. The combustion lasted about a minute. The air was dilated by the heat, but soon afterward the mercury rose above its initial level. The most important difference between this and the previous experiment, however, was that dry flakes collected on the walls of the jar. They consisted, he was sure, of phosphoric acid in concrete form. It did not absorb humidity and run into droplets as it did over water. That difference gave him hope that he would finally be able to measure the weight gained by the phosphorus in the combustion:

> It would be very interesting to collect that acid in this state in order to weigh it and determine the augmentation of the weight to see if it would be proportional to the absorption. But because this acid is very deliquescent it is to be feared that it will attract humidity from the air even while it is being transferred. To avoid part of that inconvenience I thought that it would be suitable to heat the mercury and the jar in order to transfer [the acid] hot.

Lavoisier put the apparatus over the fire, but his plan failed. The flakes of phosphoric acid melted. He could not tell whether the cause was the heat or the humidity released by the heat from the mercury.[9] In any case, he was once again unable to measure the weight gain due to the fixation of air in the combustion of phosphorus.

On June 30 Lavoisier placed 1 gros, 55 grains of foliated earth of tarter under a jar inverted over mercury. He took precautions so that the deliquescent salt would not have time to absorb humidity during the weighing. After sucking through a hole in the jar to raise the level of the mercury, he waited to see if the salt would absorb air. It is not evident why he chose to study this salt, or why he expected an absorption. Foliated earth of tarter was actually a neutral salt formed by saturating fixed vegetable alkali with acid of vinegar (also sometimes called acetous acid). Black had mentioned it briefly as one of the acids that drive off fixed air from an alkali only in the later stages of combination with the alkali. According to his general theory, the alkali should not reabsorb fixed air unless it was separated again from the acid. The salt could, in fact, easily be decomposed by heat, but Lavoisier did not heat his. Nevertheless, he did observe, by 5 o'clock the next afternoon, that the mercury had ascended nearly two lines on the jar. He seems to have drawn no conclusions about the cause of this small effect.[10] The most significant aspect of the experiment may be the indication that he had found the substitution of mercury for water covered with oil advantageous enough to begin using it regularly, at least for operations requiring only his smaller inverted jars.

By the beginning of July, Lavoisier decided that he had gathered enough experimental evidence to begin to make good on his commitment to present the details of his experiments at the regular weekly meetings of the Academy. He planned to begin, on July 3, with those concerning calcareous earth and lime. In preparing for his talk, he probably noticed some gaps in his treatment of the subject, which led him to conduct a series of experiments on quicklime and slaked lime between June 30 and July 2.[11]

Black had pointed out that when quicklime "is exposed to the open air, it absorbs the particles of water and of fixed air which come within its sphere of attraction . . . if it be exposed for a sufficient length of time, the whole of it is gradually saturated with air, to which the water gradually yields its place."[12] To determine how much weight quicklime gains in this process, Lavoisier placed a piece weighing a little over 12 ounces in a capsule containing "free air," on the morning of June 30. To measure the air absorbed, he placed a second piece under a jar inverted over mercury. By 6 o'clock the next evening, the quicklime in the capsule had already gained 4 gros. By July 2, the mercury over which the second piece had been placed had risen about 2 lines, but since the temperature in the laboratory was about 3/4 degree less than when he began the experiment, he could not be certain that the volume had really decreased. He let both experiments go on for nearly a month.[13]

To test Black's statement that quicklime absorbs both water and fixed air, Lavoisier performed a third experiment on June 30. He poured distilled water over a piece of quicklime weighing 5 gros, 22 grains, but was unable to dissolve it completely. He then bubbled fixed air through the basin, and formed a precipitate. He was, however, unable to "remove the odor of lime water from the water," even though he had employed "much acid and alkali." After converting the remaining lime to calcareous earth by means of soda, he inferred that to do it in the way he had intended, "still more fixed air would have been required."[14] Apparently, however, he did not consider the experiment important enough to repeat.

On July 2, the day before he was scheduled to present the first of his papers to the Academy, Lavoisier decided to make a measurement of the quantity of fixed air in soda parallel to the measurement of the fixed air in calcareous earth that he had determined by his experiments on the action of nitrous acid on chalk. This time he performed both types of experiment in one day: an indirect measurement by weighing the quantities of water, acid, and alkali at the beginning of the operation, and subtracting from it the liquid products of the operation to determine the fixed air driven off; and a direct measurement of the fixed air disengaged by the operation carried out in a closed vessel.

The first experiment Lavoisier summarized in a compact balance sheet (table 4).

TABLE 4

I took a small matras which weighed	4 ounce	3 gros	2 1/2 grains
I put in nitrous acid	4	—	—
Water	7	1	66
Soda in dry crystals	4	1	34 1/2
	19	6	31
After the operation, estimating a few grains for the fluid lost, I find for the weight of the material	19	0	68
Loss during the effervescence		5	35

He remarked that "the quantity of 4 ounces, 1 gros, 34 1/2 grains of soda was not sufficient for the saturation. It ought to have been around 4 ounces, 3 gros. But that matters little."[15]

For the direct measurement of the fixed air Lavoisier again had recourse to his large glass vessel, and to his device for tipping the nitrous acid into the alkali by cord and pulley. Not able to use mercury because of the large volume of the vessel and basin, he relied on his old method of spreading oil over water. Employing again just one-quarter of the quantities of alkali and acid that he had used in the open air operation, he pumped the water up, syphoned in the oil, pulled the cord, and "made the combination." The effervescence was "quite lively." All went well. The quantity of air produced remained steady at "5 inches, 10 1/2 lines on the large recipient, which gives 162.81 cubic inches." Multiplying by four to make the result comparable with the measurements made of the other components in the open air experiment gave him 661 cubic inches. To convert this volume to weight, he noted that "as the barometer is at 28 inches, 1 line and the thermometer is at 15

1/2 degrees, one cannot deceive oneself by much in estimating the cubic inch [of air] at 0.46." On this supposition, he calculated that the weight of the air was 4 gros, 11 grains. Comparing that figure with the 5 gros, 35 grains lost in the operation in the open air, representing the indirect calculation of the fixed air, Lavoisier concluded that "the excess is 1 gros, 23 1/2 grains, which is of little importance." Noticing afterward that he had made an error in measuring the vertical movement of the water level, and that it had actually been 4 inches, 10 1/2 lines, Lavoisier corrected all his calculations and came out with an "excess" of 2 gros, 2.42 grains, but this larger figure did not cause him to reconsider his judgment that it was unimportant.[16]

These two experiments and his evaluation of the results give a significant glimpse of Lavoisier's understanding of his quantitative methods on the eve of his first public presentation of detailed examples of his experiments. It is clear that he sought reliability, but not great precision. He routinely estimated the magnitude of errors due to small losses that he could not measure. He aimed for a complete balance sheet of all the materials before and after an operation, but he did not expect to arrive at measured quantities that added up exactly. If they came close enough to support his interpretation of the operation he was studying, that was good enough for him. In this case, his confidence that his interpretation was supported by his numbers was made easier by the fact that his interpretation only confirmed the interpretations that Joseph Black had already given for these processes. In converting volumes of air to weights, he was aware that changes of temperature and barometric pressure could interfere, but he was content to note that they were close enough to standard conditions so that he did not have to correct for them. He still treated the weight of fixed air as close enough to ordinary air to use the known figure for the density of the latter in his calculation. With all these compromises between the ideal and the attainable, he was forging, piece by piece, an experimental method suitable to his goals.

Even on the day of his presentation, Lavoisier began another experiment. His aim was to "pass the air of soda into lime." He was essentially returning, with the experience gained over six months, to the first of Black's experiments that he had sought, without great success, to reproduce in February (see chap. 3, p. 18). Now he was not only able to handle the qualitative operation more skillfully, but he knew, from the experiments completed the day before, how much fixed air the soda contained to pass to the lime. He dissolved 26 grains of dry soda crystals in 2 ounces of distilled water, then added to the solution 2 gros of desiccated slaked lime. After agitating the mixture, he let it stand for a while, decanted the liquid, and put it aside until morning,[17] while he left for the meeting of the Academy.

Lavoisier entitled his talk on July 3, "The existence of an elastic fluid fixed in calcareous earths, and the phenomena that result from its absence in lime." He introduced his own experiments as a repetition of the principal experiments of Black, Meyer, and their respective followers, together with some new ones that he had added. His objective, he said, was to evaluate, as far as possible, their respective systems. All of the experiments he would report, he emphasized, were "tied very exactly to one another."[18] The order in which he presented them bore little similarity to the temporal order in which he had performed them. The sequence displayed, instead, the logical order in which an investigator might perform them if he were to proceed more systematically than scientists feeling their way along ever do.

Beginning with the solution of chalk in nitrous acid, he proceeded to the measurement of the "elastic fluid" disengaged from the chalk during the process. He moved then to the determination of the "quantity of water necessary to saturate a given quantity of quicklime," and to the "extinction" of quicklime in a vacuum, which proved to be no different from the process in the open air. Next he described the solution of slaked lime in nitrous acid and the quantity of elastic fluid disengaged in that process.[19] At this point, he paused to draw some "general consequences" from the six experiments so far presented. The argument that he built on the aggregate results of his experiments is so revealing of the style of quantitative reasoning that was beginning to emerge from his investigation that it is worth quoting at length:

> First, it is evident from experiment III that the quantity of 1 ounce, 5 gros, 36 grains of slaked lime employed in experiment V, and necessary to saturate 6 ounces of nitrous acid, contained 3 gros, 0 1/4 grains of water. 2) according to experiment VI, that this same quantity of slaked lime contained 16 1/2 grains of elastic fluid; it really contained, therefore, only 1 ounce, 2 gros, 18 3/4 grains of alkaline earth: but, according to experiment I, 2 ounces, 3 gros, 36 grains of calcareous earth were required to saturate a corresponding quantity, that is, 6 ounces, of nitrous acid; from which it appears that one can conclude that 2 ounces, 3 gros, 36 grains of chalk equally contain no more than 1 ounce, 2 gros, 18 grains of alkaline earth; that they contain in addition 3 gros, 0 1/4

> grains of water, and 6 gros, 16 1/2 grains of elastic fluid. According to experiment V, these 6 gros, 16 grains correspond to 800 cubic inches, from which it follows that the elastic fluid contained in the calcareous earth weighs 561/1,000 of a grain per cubic foot, at a temperature of 17 degrees on the thermometer of M. Réaumur, that is, a little more than a half of a grain; whereas a cubic inch of common air, at the same temperature, weighs, according to the results of M. de Luc, only 455/1,000 of a grain, that is, a little less than a half grain. This difference derives either from the fact that the elastic fluid disengaged from chalk is really a little heavier than that of the atmosphere, or that it is charged with vapors when escaping from the chalk, or finally, that chalk contains more water than slaked lime.[20]

There are several remarkable features of Lavoisier's reasoning. The most familiar, because it became a stable feature of his quantitative style, was that in presenting his results publicly, he ignored the sources of error he had observed, and the approximations that he had made in his experiments, and presented the numbers resulting from his calculations as though they were exact. In this case the consequence was a striking reversal of his view about the density of the air fixed in these substances. Until now he had assumed that the weight of fixed air could be equated with that of common air, and that he could use de Luc's value for the latter as the basis for converting volumes to weights. The discrepancies that resulted in his balances he had accepted as within the range of the errors expected in his measurements. Now, supposing instead that his measured results were accurate, he used them as a basis for calculating a new value for the density of the fixed air.

But was this substance any longer, in Lavoisier's mind, really fixed air? At the beginning of his investigative program in February he had focused his attention on the properties and operations through which fixed air can be passed from one body to another. In the announcement of his new theory in April he had proclaimed with assurance that fixed air was absorbed in calcinations and combustions, and released in the reduction of calxes. Yet here, even in discussing the very operations on which Black had based his identification of fixed air, Lavoisier veiled that identification in the generic phrase "elastic fluid"! Why had he suddenly evinced such caution about the nature of the substance even as he had something new to add about its physical properties? The problem derived, for Lavoisier, not from his efforts to replicate and evaluate the phenomena Black had studied, but from his own extension of the role of fixed air to account for the phenomena of calcination, combustion, and reduction. There he had, as we have seen, run repeatedly into observations that did not fit his conception that he was dealing with fixed air. He did not see how to revise his theory to account for these anomalies. In the midst of his uncertainty, he covered his confusion by adopting a term to describe his experimental results which deliberately avoided any commitment to a coherent theoretical structure.

In the second half of his discussion, Lavoisier presented the "complementary proofs" of the composition of calcareous earth and of lime: that is, the operations in which he had produced the former by combining the elastic fluid with the latter. Here he was conforming to the customary standard of eighteenth-century chemistry, that composition was to be demonstrated by decomposing a substance and then recomposing it from the products of the decomposition. He began this section with an experiment in which he had converted lime to calcareous earth by bubbling through it air disengaged from the action of vitriolic acid on chalk. As further proof he included the measurement of the increase in density of the liquid, measured by his areometer.[21] The general conclusions he reached from his experiments on the calcareous earth and lime were, implicitly, all restatements of Black's conclusions, but without Black's identification of the substance in question as fixed air. The subtlety with which Lavoisier evaded that identification can be seen in his first conclusion:

> First, that there exists in calcareous stones and earths an elastic fluid, a species of air in a fixed state, and that this air, when it has regained its elasticity, possesses the principal physical properties of the air.[22]

Whether subconsciously or by design, Lavoisier was playing here on an ambiguity rooted in the different meanings that Stephen Hales and Joseph Black had given to the same term, fixed air.

The experiments with soda that Lavoisier began just before delivering his paper were the beginning of an effort to provide, for a fixed alkali, a set of experiments involving the fixation and transfer of the elastic fluid parallel to the set he was organizing on calcareous earth and lime for his presentation of that subject. Because the operations on lime and calcareous earth that he had performed at different times provided, when brought together, a model for his treatment of alkali, he could proceed rapidly and systematically with the latter.

The morning after his talk, July 4, Lavoisier continued from the point where he had broken off his experiment on the passage of the elastic fluid from soda to slaked lime. He placed the solid material resulting from mixing them in a bottle to combine it with nitrous acid in the "ordinary apparatus"; that is, in the large inverted vessel with the platform device for pouring the acid into the solution inside the vessel. The resulting effervescence was not very lively, because, as he assumed, the material had been soaked so long in water. The water level in the pneumatic vessel descended 3 inches, 6 lines, corresponding to a volume of 83.13 cubic inches. From his previous experiment mixing soda with nitrous acid, he estimated that the alkali had contained 135 cubic inches, and from an experiment made on June 7 mixing slaked lime with nitrous acid, that the lime contained 6 cubic inches, for a total of 141.10. In the present experiment, therefore, he could not have disengaged all of the air contained in the two substances mixed. Consequently, after dismantling the apparatus, he set it up for another experiment with the supernatant fluid left from the previous experiment, which he assumed must contain the alkali made caustic by the operation. When this fluid was mixed with nitrous acid, 64.66 cubic inches of air were released. The total of that and the 83.13 inches released in the preceding experiment was 12 2/3 cubic inches *more* than he had calculated the materials could contain. The excess could only have derived, he inferred, from "air attracted by the caustic alkali during the night."[23]

Dissatisfied, Lavoisier repeated the experiment later the same day, mixing the same quantity of soda (1 ounce, 26 5/8 grains) with twice as much (4 gros) of desiccated slaked lime, in two ounces of water. This time he obtained a large pulpy mass, from which he could decant only one-fifth of the volume of clear liquid he supposed it to contain. After washing the lime several times and allowing it to drain, he mixed it with nitrous acid in his apparatus, and obtained 133.94 cubic inches of air. Repeating the operation with the supernatant liquid yielded only 4 cubic inches. He multiplied that result by 5 to give 20 cubic inches as the amount he would have obtained if he had been able to separate all of the liquid. The total of air that should have been released from the lime and the alkali contained in the liquid was, therefore, 153.94 cubic inches, of which 12 cubic inches had, according to the experiment of June 7, been already contained in the slaked lime before the operation. The quantity of air "in the machine" which had come from the soda, 141.94 cubic inches, still exceeded, "by 7 inches of air" the 135.10 cubic inches that the experiment of July 2 had shown to be disengaged by the action of nitrous acid on the same quantity of soda.[24] That was only a little more than half as much as he had had to attribute in the preceding experiment to absorption of air by the alkali after the operation.

His interpretation of this set of experiments, together with the reasoning about the set of experiments on calcareous earth and lime in the paper he had just delivered at the Academy, reveal how quickly Lavoisier was coming to perceive the extended potential of his quantitative method, now that he had grasped the fundamental nature of the weight relationships. Even though he did not convert volumes to weights in this case, there was no confusion between weight and volume relations. He needed only to account for the transfers of an elastic fluid whose density he assumed to be constant through the operations. Therefore, he was reasoning in terms of quantities conserved in the transfers. His complete confidence in that conservation allowed him to construct interpretations of extended chains of chemical changes drawing on results obtained from multiple experiments. None of his experiments had yielded accurate enough balances between the materials present before and after an operation to prove the validity of his assumption. He never gave any sign that he thought that the principle needed demonstration. On the contrary, it became now so basic to his way of thinking about chemical changes that he made whatever ad hoc hypotheses he needed to fit the results of his experiments to the principle.

After completing this experiment, Lavoisier was either satisfied that he had sufficient evidence to support a discussion of the elastic fluid in alkalis at his next presentation, or he was interrupted by some other affairs. After ten days in which he carried out no experiments, except for three aerometer measurements of water from a limestone pit,[25] Lavoisier took up, on July 14, a part of his broader investigative project that had lain largely fallow since his Easter memoir.

In his discussion of the dissolution of metals in a number of menstrua as another means to combine them with fixed air, in his Easter memoir, Lavoisier had pointed out that, if one afterward precipitated the metal by joining an alkali to the acid with which one had dissolved the metal, "the phenomena are very different according to the state of the alkali employed." If the alkali were in its "natural state," that is, if it contained fixed air, then "the precipitate is much heavier than the metal had been in solution." If the alkali were caustic, on the other hand, deprived of its air, then the "augmentation of weight of the precipitate is almost null, the reduction is easier, the metal contains much less air, . . . and it precipitates with a color approaching its metallic color."[26] Although he presented these generalizations as observations, there is no evidence that he had performed the experiments to support them. Most likely they were implicit predictions arrived at by connecting his theory with Black's theory of causticity. After one unsuccessful start on May 6 (see chap. 6, pp. 48–49), he had not further explored these phenomena. Three days before the meeting of the Academy on July 17 at which he planned to present the next installment of his experimental researches, he decided that it was urgent to include in his paper an account of such experiments, and he set out to do them.

Lavoisier began on July 14 by dissolving mercury in nitrous acid. Nitrous acid was long known to dissolve mercury and other metals very easily, emitting heat and abundant red vapors.[27] Lavoisier put 12 ounces of mercury, together with 12 ounces of nitrous acid, into a matras with a narrow neck. The lively effervescence heated the matras and filled it with red vapors. From his examination of Priestley's publications, Lavoisier knew by this time that the vapors constituted the new "species" of air that Priestley called "nitrous air."[28] Consequently, Lavoisier expected the escape of the vapors to produce some loss of weight, and placed the matras on his balance even before the dissolution was complete. He observed that there had been "only 1 gros, 18 grains of loss." Three hours later, when no undissolved mercury remained, and he weighed the matras again, "I was very surprised to see that the loss of weight was no more than 54 grains." The solution had gained 36 grains since the first weighing. (1 gros, 18 grains = 90 grains. −90 + 36 = −54.) The next morning, when he estimated that there were 18 grains of metallic mercury left, the overall weight loss had decreased still further, to 18 grains. Consequently, to compensate for the initial loss of 1 gros, 18 grains, "the solution must have gained 1 gros in weight." Unable to account for the source of the gain, and too pressed for time to pursue the phenomenon further, Lavoisier added distilled water to prevent the solution from crystallizing, and divided it into six bottles, four of which he used for the precipitations with different forms of alkali.[29]

The four precipitations Lavoisier planned were with chalk, desiccated slaked lime (i.e., with the ordinary and caustic alkalis of calcareous earth), soda, and caustic soda. Of each alkali he added to one of the bottles an amount calculated to saturate the 2 ounces of nitrous acid that each bottle should contain. The bottle into which he placed 6 gros, 60 grains of chalk effervesced throughout the time that the precipitate formed. He collected the yellow precipitate, decanted the liquid from it, and dried it. The precipitate weighed 2 ounces, 4 gros, 60 grains, a gain of 4 gros, 60 grains over the 2 ounces of mercury dissolved in the solution. The mercury precipitated by desiccated slaked lime was deep gray. When dried it weighed 2 ounces, 2 gros, 54 grains. That produced by the soda was "a quite beautiful yellow." He dried it, but did not record its weight. The caustic soda precipitated without effervescence a deeper yellow substance, which weighed 2 ounces, 2 gros, 36 grains.[30]

The weight gained by the mercury precipitated with chalk was favorable to Lavoisier's belief that its source was the elastic fluid contained in that alkali, but the fact that the precipitate formed by slaked lime had gained more than half as much as that formed by chalk did not fit his view that precipitates made by caustic alkali should cause little or no augmentation. Thinking that he may not have dried the precipitates sufficiently to remove all of the mother liquor they had imbibed, he washed them several times in distilled water and dried them thoroughly. The precipitate by chalk now appeared very dry, and weighed 2 ounces, 2 gros, 69 grains. The precipitate by slaked lime weighed 2 ounces, 1 gros, 5 grains, but it looked to him still "a little less dry," and he proposed to weigh it again. There is no record that he did. He dried the precipitate by caustic soda further, but accidentally lost it, which "prevented" him from judging "the effects of the desiccation."[31]

On the very day that he was scheduled to present his memoir, Lavoisier made one more attempt to improve on these results. Having kept the solution of iron in nitrous acid that he had produced on May 6, in his abortive first investigation of these phenomena, he used that solution on July 17, to repeat on iron the four precipitations with chalk, slaked lime, caustic alkali, and soda that he had performed on the mercury solution. He separated his solutions into portions, each containing 2 ounces of nitrous acid and about 2 1/2 gros of iron. The yellow to reddish gray substance produced by drying on a sand bath the precipitate formed by chalk weighed 6 gros, 59 grains. A precipitate of the same color produced by the slaked lime weighed 4 gros, 69 grains. Caustic alkali produced, without effervescence, another similarly colored precipitate weighing 5 gros, 4 grains, whereas soda produced, with "swelling and effervescence," only 4 gros, 44 grains.[32] Taken altogether, these results hardly clarified the picture. As we shall see, however, Lavoisier picked out from them what he could best use at the moment, and left the problematic results to deal with at some future time.

Remarkably, Lavoisier even made one more attempt, on the day of his presentation, to detonate niter under a jar with the burning lens. This was another of the processes he had invoked in his Easter memoir as a means to fix air in metals, but had not succeeded in producing in a closed vessel where he could measure the absorption. He mixed 4 gros of melted, desiccated niter, passed through a silk sieve, with 1 gros of braised charcoal also passed through the sieve, and mixed them well on a crystalline mortar. Placing them in a varnished earthen capsule, he covered the mixture with a large inverted vessel. The detonation took place, lasting less than half a minute. The water level quickly descended 16 1/2 inches, but soon began to rise again.[33] At the time he delivered his paper at the Academy he did not yet know whether in the end there would be any absorption.

His second account to his colleagues of his research on the elastic fluid fixed in various substances, Lavoisier devoted, with one exception, entirely to the experiments he had conducted during the immediately preceding two weeks. In the first part he described the set he had performed between July 2 and July 5, summarized under the heading "The existence of an elastic fluid fixed in alkalis, and the means to deprive them of it." He introduced his topic by stating that, after he had proven in his previous presentation that there is an elastic fluid in calcareous earth, and absent in lime, it remained for him now to "follow the combinations of this fluid with substances of different nature, notably with alkaline substances and with metals."[34] The most interesting aspect of this statement of purpose is that, even in his cautious refusal to identify the elastic fluid as any specific air, he clung to his assumption that the same elastic fluid entered into all the combinations he was studying.

After explaining that he had chosen purified crystals of soda, rather than salt of tarter, to examine the elastic fluid in fixed alkalis, because the hygroscopic property of the latter made it difficult to determine its dry weight, Lavoisier launched directly into detailed descriptions of his experiments. The centerpieces were the paired experiments on the dissolution of soda in nitrous acid that he had carried out on July 2 (see above, pp. 87–89). To facilitate comparisons with the parallel experiments on calcareous earth that he had reported in his previous paper, however, he multiplied the measured weights from the first of these experiments by 1 1/2, to correspond to the quantity of nitrous acid (6 ounces) used in the calcareous earth experiment. He then multiplied the volume of air measured in the second soda experiment (135 cubic inches) by 6 to give the volume (810 cubic inches) that he would have received, if he "had used the same doses" as in the preceding experiment. Now, however, he transformed his interpretation of these experiments in the same way that he had previously done with the calcareous earth experiments. Where he had initially converted the volume of air disengaged from the soda to a weight with de Luc's figure for the weight of common air, and regarded the resulting difference from the quantity arrived at from the loss of material in the first experiment as unimportant, he now treated that difference (3 gros, 3 5/6 grains when scaled up) as evidence that the air disengaged by the effervescence was "heavier than the air of the atmosphere, or that it had carried water vapor off with it."[35]

Lavoisier built further inferences on the comparison between these experiments with soda and the earlier ones with chalk. The fact that it had required "much more" soda than chalk to saturate the same quantity of nitrous acid, "indicates," he claimed, that the composition of the former includes much water of crystallization. The quantities of elastic fluid that the two substances contained were "sufficiently exactly proportional" to their respective quantities of "alkaline substance." In saturating the same quantity of nitrous acid, the chalk disengaged 800 cubic inches of air, the soda 810: quantities that "can be regarded as sensibly the same."[36]

Extending this reasoning, Lavoisier noted that from the measured quantities of chalk and soda required to saturate the nitrous acid, one could "make a calculation, sufficiently probable, of the proportion of water, elastic fluid, and alkaline substance, that the soda contains." Although he acknowledged that it would require a few more experiments to "give this calculation a certain degree of evidence," he offered one anyway, and claimed that the "calculation cannot be far off from the truth."[37] Was Lavoisier reasoning soundly, or was he allowing himself to be swept away by these heady possibilities for building calculation on calculation, starting from values based on single experiments which he knew to be imperfect? It is easy, in retrospect, to identify the risky features of his claims, but necessary also to remind ourselves how original and venturesome his approach had become. By now, Lavoisier was moving beyond the methods of Hales and Black that had served as his models, to a form of quantitative experimentation and reasoning that no one before him had practiced. It is not surprising that, in these brief weeks during the spring and summer of 1773 when his intense burst of experimental activity carried him to this point, he did not yet know the limits of his methods.

Next Lavoisier inserted the experiment, conducted on June 4, in which he had followed with his areometer the diminution of the specific gravity of alkali after successive additions of slaked lime (see chap. 7, pp. 74–75). His account was a close paraphrase of the record of the experiment as recorded in his notebook, but he added the information that after the third dose of lime, further additions did not change the specific gravity, and claimed on this basis to be able to determine the proportions of slaked lime to soda (3 parts to 2) necessary "to bring the soda to the state of causticity." He had originally performed this experiment, as we have seen, to "decide at one stroke," between the "systems" of Meyer and of Black. Publicly, however, he hedged. "As favorable as this experiment appears to the system of Black," he said, "it could, nevertheless, still be explained in that of Meyer." His partisans could say that the lime had furnished to the alkali a "material lighter than the solution," just as the addition of a spirituous liquid to water lessens its specific gravity. "It would even be probable that this matter was nothing other than phlogiston; finally, that they would add that this property of phlogiston to diminish the specific gravity of liquids in which it is combined is an effect well known in chemistry." After offering this possible explanation to the partisans of Meyer, Lavoiser quickly withdrew from his own suggestion. "I shall not stop here to discuss these objections," he declared, "it is to experiments alone that one must have recourse."[38] Was this a weak effort to appear open-minded, or an ironic gesture to partisans of Meyer who might have been present? That this experiment, considered in isolation, could be interpreted in two ways, suggests only that Lavoisier had been rather impulsive in thinking at the time he performed it that it could, by itself, settle the question. Since then, he had conducted numerous experiments whose design and interpretation presupposed his acceptance of the system of Black. To invoke phlogiston as a possible agent for the diminution of the density of the solution was a strange ploy for someone who had, in his historical review, depicted Meyer's entire system as "directly contrary" to the phlogiston theory of Stahl.[39] It is hard to avoid the impression that Lavoisier's purpose in raising this issue was mainly to flaunt his own freedom from theoretical systems and his devotion to experimentation.

The two experiments on the passage of air from soda to lime that he had performed on July 3 and 4, Lavoisier reported, for the most part, as he had recorded them in his notebook. Through a combination of a small readjustment in one of his measurements, and perhaps a realization that in comparing the total of the airs released by the alkali and the lime to the quantity contained in soda he had omitted to include also the air contained in lime, he reduced the discrepancy from 12 1/2 to 3 cubic inches, obviating the necessity to infer, as he had initially, that the alkali had absorbed air overnight (see above, p. 93). The difference now looked to be as little as could be expected for experiments of this type.[40]

Concluding the first section of his paper, Lavoisier asserted that

> it has been proven, almost as much as can be in physics, that the same elastic fluid that had been recognized in calcareous earth, [as discussed in the memoir he had read two weeks earlier,] . . . exists equally in alkalis . . . that it can be displaced by dissolution in acids, and that the effervescence that can be observed in the moment of combination is an effect of the disengagement of this fluid.[41]

He added that it had also been shown that this fluid has more affinity for lime than for saline alkalis. He did not remind his listeners again that what he believed he had proven so conclusively was a subset of the same phenomena for which Joseph Black had already given powerful experimental evidence seventeen years earlier. His attitude toward Black's achievement is best revealed in a note Lavoisier later added, before publication, to the summary of Black's work that he had included in the historical memoirs he had read to the Academy during the spring: "The theory of fixed air as it left the hands of Black had not acquired the coherence and consistency that one [i.e., Lavoisier himself] has given it in this article." Those qualities had been given to it by Nicolas Jaquin, in his examination of the doctrines of Meyer and of Black published in 1769. He did not make this comment, Lavoisier added, to diminish the "admiration due to the genius of M. Black, to whom is due the merit of the invention," but to render justice also to the contribution of Jacquin.[42] These sentiments could also be applied to Lavoisier's probable opinion of his own contribution. Without diminishing his admiration for Black, he might have felt, that system now owed its strongest proof to his own efforts.

In the second part of his paper, Lavoisier turned to "the combination of the elastic fluid of calcareous earth and alkalis with metallic substances by precipitation." Having performed all of the experiments he described in this section during the last three days before he was to present them, he must have had very little time to write them up. His discussion, in fact, bears the marks of his haste, as well as of the unfinished state of the work that he wished, nevertheless, to make public.

Lavoisier introduced this topic with a bravado echoing the style in which he had introduced his "new theory" in April:

> A sufficiently large number of experiments lead me to believe that the elastic fluid, the same one whose existence I have searched to prove in calcareous earth and in the alkalis, is capable of combining with the majority of metallic substances by precipitation; that it is in large part this principle which provides the augmentation of the weight of metallic precipitates, which removes their lustre, which reduces them to the state of a calx, etc. Although my experiments on this subject have already been multiplied, nevertheless, because one cannot doubt that the precipitates retain something with them, and their dissolvents, and materials employed to precipitate them; that to this circumstance are joined still other particular phenomena caused by the decomposition of the acids; I believed that I should reserve for a special memoir the greater part of my experiments. Consequently I give here [only] those that are connected most essentially to the subject that I am treating today, while alerting the listener, that I give them only as facts whose consequences are not yet sufficiently proven.[43]

Among other things that Lavoisier was learning well, was how to justify eloquently the premature announcement of incomplete investigations.

The "sufficiently large number of experiments" consisted, as far as can be told from Lavoisier's laboratory record, of the four precipitations of mercury, respectively, with chalk, lime, soda, and caustic alkali: and the parallel set of precipitations of iron with the same alkaline substances (see above, pp. 95–97). The "multiple" experiments he chose not to report were the precipitations of mercury and of iron with soda and with caustic alkali.

Lavoisier interpreted these results as favorable to his belief that an elastic fluid "contributes" to the augmentation of the weight of the precipitates of mercury. Calcareous earth, containing fixed air, increased the weight more than did slaked lime. He got around the awkward fact that, according to his theory a caustic alkali should add no weight to the metal, by assuming that his slaked lime "still contained some portions of elastic fluid."[44] He avoided the still more awkward fact that the precipitate of iron with caustic soda was heavier than that with ordinary soda, by not discussing those results. As in April, he justified his avoidance of full disclosure by promising that "I will occupy myself some day more particularly with this subject."[45]

Lavoisier's assertion, at the end of the first section of his paper, that the phenomena he described had been proven "as much as is possible in *physics*" is one among a number of similar statements that appear to support the argument of recent historians that he sought throughout his career to apply the methods of experimental physics to solve the problems of chemistry.[46] To assess the meaning of such references, however, it is essential to connect each of his statements to the context in which Lavoisier made it. Having followed him through the investigation by which he believed he had delivered this proof, we may ask, which of the methods that he had used did he identify with physics? The most obvious answer is his application of the areometer to determine the densities of solutions. In his discussion of this instrument in 1768, he had written that "the physicists [*les physiciens*] have imagined different means to determine the density of fluids," and he advised readers to look up descriptions of the instrument and its uses in "physics books [*les livres de physique*]."[47] But these measurements played only a subordinate role in this investigation. In the majority of his experiments, including the most decisive of them, he used a combination of methods similar to those introduced separately by Joseph Black and Stephen Hales. Black was trained as a physician. The methods he used in the investigation of fixed air were drawn mainly from mainstream mid-eighteenth-century chemistry. Hales was a rural pastor, heavily influenced by Newton, who used quantitative methods to investigate phenomena ranging from plant and animal physiology to the fixation of air in a multitude of bodies. Through the mediation of Lavoisier's chemistry teacher, Guillaume-François Rouelle, the pneumatic vessel introduced by Hales was coming, by the time Lavoisier entered science, to be a common apparatus in chemical laboratories. If Lavoisier regarded the methods he had applied, collectively, as drawn from "physics," then he must have had in mind something closer to what we mean today by the "physical sciences" than a specific domain associated with a well-bounded discipline. Other chemists, including Pierre Joseph Macquer, sometimes used the term "*physique*," or "*saine physique*" in a broadly generic sense to imply that rigorous methods had been used. Lavoisier may well have employed the word as his peers used it, without giving extensive thought to what he meant by it. He was pragmatic enough to be less concerned with the heritage of his methods than with their effectiveness for his purposes.

* CHAPTER NINE *

The End of the Beginning

IN THE memoir that he read to the Academy on July 17, Lavoisier remarked, as he made the transition from the first half of his paper on the elastic fluid contained in alkalis, to the second half on metallic precipitations:

> Perhaps this would be the moment to report the experiments that I have made on the nature of the elastic fluid disengaged from saline alkalis and alkaline earths; nevertheless, other considerations oblige me to occupy myself first with the combination of this same fluid with metallic substances.[1]

There is a poignancy in the way this eager young scientist approached, and then detoured around, a public discussion of the nature of the elastic fluid he so yearned to identify, but which still so eluded his grasp that he could give it no name.

Aiming to complete his presentation of the results of his research on this elastic fluid at the meeting of the Academy two weeks later, Lavoisier rushed onward in a final attempt to resolve his outstanding problems. On July 19, he extended his study of the action of lime on alkalis to volatile alkali. In doing so he was still following the program laid out by Black, who had reported briefly that he had "also rendered the volatile alkali caustic" with lime and "obtained an exceedingly volatile and acrid spirit which neither effervesced with acids, nor altered the transparency of lime-water."[2] Lavoisier's own first efforts to deal with volatile alkali, in February, had, as we have seen, only raised difficulties for his theory (see chap. 3, pp. 19–20). Now after his success dealing with one of the two fixed alkalis, he could return with more potent methodological resources to volatile alkali.

Lavoisier began by repeating, with volatile alkali, the areometer measurements of the diminution of weight after successive additions of lime that had succeeded so brilliantly with soda. Dissolving 2 ounces of concrete volatile alkali in 18 ounces of water, he plunged his silver areometer into the limpid solution and found that he had to add 3 gros, 1 3/4 grains to the instrument to sink it to the mark on its stem (for a total weight of 9 ounces, 3 gros, 65.75 grains). He then added 1 ounce of slaked lime, agitated the mixture, and waited for the lime (now presumably calcareous earth) to fall to the bottom. When the liquid had cleared, he decanted it and again plunged the areometer into it. Now he had to add only 1 gros, 66 grains to reach the mark (total weight: 9 ounces, 2 gros, 59 grains). Repeating the operation with 4 gros of lime, he had this time to add only 65 grains to the areometer, and after a further addition of 4 gros of lime, merely 5 grains.[3]

Once more Lavoisier added 4 gros of lime, but this time, "as my silver pese liqueur was not made for liquids lighter than water, it sank to the bottom." Consequently he had to make a glass pese liqueur of the Fahrenheit design. Testing the new instrument first in distilled water, he found it required 1 gros, 4 grains to sink to its mark on the stem. In the decanted liquid resulting from his last addition of lime it required slightly less weight, 66 grains. After one more addition of 4 gros of lime, the pese liqueur needed 66 1/2 grains. In addition, the liquid was very slow to clear. Lavoisier concluded that he had now added more lime than necessary to saturate the volatile alkali. He then collected the lime precipitated from all of the additions, and found it to weigh 3 ounces, 4 gros, 36 grains, a gain in weight of 4 gros, 36 grains over the total of 3 ounces of slaked lime that he had added.[4] The experiment confirmed, therefore, for volatile alkali what he had previously established for fixed alkali.

Instead of pursuing this success by repeating immediately the set of experiments by which he had measured the elastic fluid passed from fixed alkali to lime, Lavoisier first turned back, on July 20, to the combustion of phosphorus. He seemed to prefer, in the time he had left, to try to make some progress on all fronts of his research campaign rather than to concentrate all his effort on one of them. In the case of phosphorus, he still hoped to find some means to overcome the obstacle to his measurement of the weight gained in the combustion posed by the rapid absorption of moisture.

In a little glass capsule placed within a bottle, Lavoisier put a small amount of phosphorus. He recorded the weights of the three objects as shown in table 5.[5] In his previous experiments on the combustion of phosphorus he had presented the same information in narrative form. In experiments modeled on those of Black, he had begun earlier to record the weights in this tabular format. The change is only stylistic, but somehow does make more transparent the "balance sheet" method. Perhaps the spread of this style to his combustion experi-

TABLE 5

The bottle weighed	5 ounces	3 gros	2 grains
The capsule	1	0	56 1/3
I added of phosphorus			6
Total of weight	6	3	64 1/3

ments can be viewed as a symbol of the spread of his awareness of the scope of a distinctive method from the special localities which had nurtured it to the broader terrain of his research landscape.

Lavoisier again carried out the combustion in a jar inverted over mercury to obtain concrete phosphoric acid, and again ignited it by means of the burning glass. This time his plan was to collect and weigh the flakes of acid so rapidly that they would not have time to absorb water. During the combustion the mercury level rose 9 to 10 lines, which corresponded to between 10 1/3 and 11 1/2 cubic inches of air absorbed, respectively.[6]

Having prepared his weights in advance, Lavoisier removed the glass jar, seized the bottle, placed it immediately on his balance, and completed the weighing, all within half a minute. The weight was, he wrote down, "found to be 6 ounces, 3 gros, 70 grains, which gives an augmentation of 6 grains." On second thought, he wrote cautiously above that weight, "at least." There was still, he then noticed, about one grain of phosphorus left in the capsule, so that the augmentation "must be attributed to the combustion of 5 grains of phosphorus." Moreover, there was at least one grain of flowers of the acid attached to the pneumatic vessel, but it was now too late to weigh them, because they had already attracted moisture. Within five or six minutes they were reduced to liquid. On the other hand, because the phosphorus that he had weighed might have attracted some humidity even during the brief time that it had taken him to transfer it to his balance, he decided to assume that the two sources of error canceled one another, and to "restrict the augmentation to 6 grains in all." He estimated that the absorption of air had been a little more than 2 cubic inches per grain of phosphorus.[7]

By the next day Lavoisier was worried that the diminution of volume "that I have attributed to the fixation of the air" might derive instead "exclusively from the humidity of the air which was absorbed, and that the air thus deprived of its humidity might have diminished in volume." It occurred to him that he could determine whether there was simply an "absorption of water, or if another substance contained in the air contributed," by performing a second combustion in a vessel saturated with water vapor. He set it up like the preceding experiment, except that he put only 5 grains of phosphorus in the capsule, and added 36 grains of water in a second capsule, covering both of them with the same glass jar "plunged into mercury." This time, when he inflamed the phosphorus with the burning lens, it formed liquid drops instead of dry flakes. Nevertheless, "the absorption of air was almost the same." As a further check on the effects of water vapor on the volume of the air, he focused the light from the lens on the capsule of water until it boiled. Vapors spread through the vessel, but "the height of the mercury did not vary." After deducting one grain for some phosphorus found afterward unburned, he estimated that the volume of air absorbed—8 to 9 cubic inches—was, just as in the last experiment, "about 2 cubic inches of air . . . for each grain of phosphorus, or even a little more. The absence or presence of water," he concluded, "neither augments nor diminishes the absorption of air."[8]

Having eliminated this potential objection to his interpretation of the combustion of phosphorus, Lavoisier could be satisfied, at least for now, that his approximate measurement of the weight gained did represent, at least in large part, air absorbed. He went on, therefore, during the same day, to two other subjects that he had previously taken up briefly.

Following "the example of Messieurs Black and Jacquin," Lavoisier wanted to move from acids, calcareous earth, alkalis, and the elastic fluid "combined three to three," to the more complicated "combinations four to four" that they had studied;[9] that is, to the precipitation by alkalis of calcareous earth dissolved in nitrous acid. He had carried out two such operations on June 8 (see chap. 7, p. 79), but had moved quickly back to the simpler combinations that were already hard enough to analyze quantitatively. With success in that area now behind him, he took up one of these "four to four" combinations again on July 21. In his previous effort he had precipitated the calcareous earth by means of caustic fixed alkali, but failed to obtain a precipitate with caustic volatile alkali. Now he was ready to make another attempt with the latter. Probably as a control, he carried out a parallel experiment using concrete (noncaustic) volatile alkali. For each test he prepared a solution containing 2 1/2 ounces of nitrous acid and "5 gros, 45 grains of slaked lime, or 8 gros, 44 grains of chalk." The solution to which he added noncaustic volatile acid gave off a "light effervescence" and formed a slightly yellowish precipitate weighing, when washed, 7 gros, 43 grains. For the second solution he used a volatile alkali "very free of fixed air," but the "precipitate, whether in a large amount of water or otherwise, was almost null." Having tried twice, he treated this negative outcome, not as a failure to duplicate the observations of Black and Jacquin, but as a decisive outcome: "One can, therefore, consider that very caustic volatile alkali does not precipitate calcareous earth."[10]

Later on the same busy day, Lavoisier made one more effort to obtain evidence for the view of metallic precipitations that he had presented in his last memoir at the Academy in spite of its shaky experimental foundations. Employing the same solution of mercury in nitrous acid that he had used in his earlier attempt, he now compared the weights of the precipitates formed by noncaustic volatile alkali, caustic volatile alkali, and soda. Each solution contained 1 ounce of mercury. The precipitate from soda weighed 7 gros, 47 grains, that from noncaustic volatile alkali 6 gros, 56 grains, and that from caustic volatile alkali 7 gros, 53 grains. These were, obviously, useless results for him, the precipitates weighing less than the mercury with which they were supposed to combine. "All of the precipitates weigh less than they should," he commented, "which made me suppose that the fire used to desiccate them was too strong and has volatilized some of the mercury or of the precipitate."[11] He did not try again. Probably resigned to having nothing new to report on this subject for his next presentation, he went on the next day to an experiment he had never before tried.

Using a Hales apparatus and a glass retort, Lavoisier attempted, on July 22, to reduce a precipitate of mercury "without addition." It had long been known that calxes made from mercury possessed an unusual property. They could be reduced to the metallic form by heating them without the charcoal necessary to reduce the calxes of other metals. There was, however, ongoing debate about whether these were true reductions. Lavoisier used a "precipitate of mercury by chalk," which was, according to his theory, a true calx consisting of mercury combined with fixed air (or, when he was being cautious, with the unnamed elastic fluid). He placed the precipitate alone in the retort. As he began to heat it, the air in the Hales vessel "dilated as usual," but toward the end of the operation the water began to rise, even though the heat had not diminished. He attributed the diminution to "a calcination of mercury"; that is, he thought that some of the mercury formed early in the operation by reduction had subsequently been recalcined. He could see only "a few atoms" of fluid mercury in the retort. The experiment ended in the same kind of disaster that had overtaken his early experiments with ordinary retorts: "The fire having melted my retort at the end of the operation, I could not tell exactly the quantity of air produced or absorbed." After breaking the retort open to examine its contents, he found only a few drops of fluid mercury, together with various other known combinations of mercury, including black and red sublimates, and turpith mineral.[12]

Why did Lavoisier interject this new experiment at a time when it was most urgent to consolidate earlier results? Did he expect that the reduction of a calx without charcoal could contribute some strategic clarification to his troubled theory of calcination, or did some less focused whim or curiosity induce him to examine this prominent anomaly? After one attempt, he dropped the subject. That does not necessarily mean that he had no sustained interest in it. Rather, with time again running out for him, he had once more to confront the deficiencies in the experiment on which his theory of calcination rested most heavily: the reduction of the calx of lead.

After the second experiment on the reduction of minium, carried out on June 28, Lavoisier had noticed some water in the apparatus, and thought that it might account for some of the difference between the weight of the air disengaged and the loss of weight of the calx (see chap. 8, pp. 83–84). Before attempting a complete new experiment, he wanted to find out how much water might be released in the operation. To do so, he placed the same mixture of 6 ounces of minium and 6 ounces of powdered charcoal that he had used before into an iron retort, but instead of connecting it to his pneumatic apparatus with the large inverted jar, he "adapted it to a ballon pierced with a small hole." The air disengaged in the process escaped through the hole, while the water he expected to collect in the receiver. At the "beginning of the operation 24 or 30 grains of phlegm" passed into the receiver. It had a slightly piquant taste, due to the fixed air with which it was impregnated. The material remaining in the retort he found afterward still to contain some charcoal. It weighed 5 ounces, 7 gros, 30 grains.[13]

Once again, Lavoisier found himself racing to complete the experiments he needed, in time to include them in the memoir that he hoped would climax his reports to the Academy of his research activities on the elastic fluid whose transfers to and from other bodies he aimed to track down. The rest of the experiments on the reduction of minium that he produced for the occasion are dated July 31, the same day that he read his paper. It is implausible, however, that he could have performed them all on that day. Most likely, he was too busy during the preceding week to transfer the immediate records he kept on separate pieces of paper during the course of the experiments to his laboratory notebook, and entered them all at once on the day he finished an extensive series of operations.

To determine how much of the 5 ounces, 7 gros, 30 grains of material found in the retort at the end of the reduction operation carried out on July 22 was charcoal, Lavoisier put this residue into an iron ladle and heated it to try to burn off the charcoal. He soon noticed that the surface of the lead was being calcined. "Nevertheless with precautions, I was able to preserve a very considerable quantity of melted lead. The rest was partly in granules, partly a calx. I did not dare go further, for fear of calcining too much of the lead." He placed the residue on the balance and found it now to weigh 5 ounces, 4 gros, 10 grains. To correct for the weight gained by the lead which had been calcined, however, he "reduced [this figure] by estimation to 5 ounces, 3 gros, 60 grains." The weight lost in this calcination after this correction, 3 gros, 52 grains, represented the charcoal present uncombined in the original residue. That left, from the original quantity of 6 ounces of charcoal employed, only 2 gros, 20 grains that could have entered into "the combination of the reduction."[14] From this result Lavoisier could also calculate the proportion of lead in minium: 5 ounces, 3 gros, 60 grains out of 6 ounces, or, 89 pounds, 10 ounces of lead per 100 pounds of minium.[15]

To elucidate further the interaction between minium and charcoal during the reduction process, Lavoisier placed four pieces of charcoal weighing 2 1/2 gros into a crucible, together with 6 ounces of minium. As he heated the crucible, he watched to see if the charcoal would "diminish." The surface of the minium became much deeper red than its usual color, perhaps, he thought, "by combination with the fixed air that is emanating from the bottom from the reduction." A little later the minium not in contact with the charcoal melted to litharge and then became glassy. The glassy substance was not reduced even when it came into contact with the charcoal, "as if the minium, thus charged with fixed air, had become irreducible." At the bottom of the crucible he found afterward 2 ounces, 2 1/2 gros of lead, or nearly half of the original weight of the minium. The charcoal seemed to have been diminished and, as if dissolved on the surface, "but it was not easy to judge."[16]

Obviously fascinated by these observations, Lavoisier noted that "this experiment merits repeating." By then, however, his most urgent business was to get on with the "new reduction of minium or lead calx" for which he was preparing. Among his preparations was to make separate measurements of the diameter of his large glass vessel at four different zones along the two and a half feet of its height, and to calculate the cubic inches of volume corresponding to inches and lines within each of the zones, so that he could make quick and accurate calculations of the volume of air released during the course of the experiment. The fifth zone, at the top of the vessel, could not be calibrated this way, because its diameter varied. He determined its volume by the quantity of water necessary to fill it.[17]

As in the experiment of May 23, Lavoisier connected an iron retort to a receiver placed under his large vessel immersed in water. He also attached his pump, a series of four bottles containing lime water, and a final bottle inverted over water to collect the air that traversed all of these bottles without being absorbed. Again he placed 6 ounces of minium with 6 gros of charcoal in the retort. Although his procedures were the same as in the earlier attempts, the operation went better. "I pressed the fire," he related, "until the point at which no air passed. The disengagement was greater than ordinary." He believed that he had "caused all of the fixed air to depart, which did not happen for me in the preceding times." This more auspicious result of Lavoisier's third try appears not to have resulted from any particular technical change, but from the general experience he had gained in managing a difficult operation. Using the calibrations of the five zones of the vessel that he had previously tabulated, he was able immediately to calculate the volume of the air released by the operation. The total was 559.9129 cubic inches, a quantity that he sensibly reported afterward as 560 cubic inches.[18]

After completing the operation, he allowed the apparatus to stand for forty-eight hours. It is possible that he carried out the experiment in the crucible, described above, during this period rather than before beginning the reduction. Several smaller experiments, including a determination of the "degree of acidity" of his nitrous acid by titrating it with quicklime, measurements of the density of water impregnated with fixed air, and of mixtures of the impregnated water with lime water, he may have carried out also while letting the air disengaged in the reduction of minium stand in the apparatus. On July 28 he recorded the last observations of the air disengaged in the fermentation experiments he had begun in mid-May (see above, chap. 6, p. 54). The volume had contracted a little more, but there was nothing else to observe.[19]

When he observed the level of the water in the vessel two days after the minium reduction experiment, it had fallen 1 inch, 2 1/2 lines.[20] Better prepared than the first time to record the relevant weights, Lavoisier filled the four absorption bottles with measured quantities of lime water. With the first stroke of the piston of his air pump, Lavoisier observed that the solution became slightly cloudy. With the second stroke it became more so, and the second bottle began to cloud. The third stroke began to affect the third bottle. It took six strokes to form the first clouds in the fourth bottle. He continued to pump air until the inverted bottle at the end of the line was filled. The marks on the calibrated large vessel now enabled him to determine immediately the volume of the air he had removed from it. The level had risen 4 inches, 9 1/4 lines in the second zone, equivalent to 126.84 cubic inches.[21]

Having previously determined by filling it with water that the volume of the bottle in which he collected this air was 63.83 cubic inches, Lavoisier "supposed the contents of the bottle to be 66 cubic inches" to allow for "the little bit of air that is always lost in operations of this kind." "One sees, therefore," he concluded, "that in passing through the lime water the 126 cubic inches were reduced in that operation to 66, which makes a reduction by half."[22] Afterward he noticed that there was "a little error in the operation, because air had escaped," and he inferred, accordingly, that there had "really been only 121 cubic inches employed."[23]

Lavoisier placed a rat into the bottle in which he had collected the air not absorbed by the limewater. It suffered sensibly. It breathed with difficulty, and fell on its side. He pulled the animal out after five minutes, and it quickly recovered. Filling another bottle with the air, he found that it was necessary to remove 124 more cubic inches from the vessel. Accordingly, about as much of the air had been absorbed by the lime as from the first portion pumped. The air in the bottle "still extinguished a candle just as fixed air would have done." He filled a third bottle, noticing that, while he did so, the first lime water bottles were beginning to clear, because, he assumed, the fixed air was dissolving the calcareous earth. He consumed this time 114 1/4 cubic inches of air from the large vessel. Into this bottle he put another rat. The animal suffered more than the first one had. "In less than one minute it fell on its side. I retrieved it, but too late. It was dead."[24]

Filling one more bottle, which required him to pump only 111 1/2 cubic inches from the large vessel, he put a mouse in it. The smaller animal "perished in a third of a minute," with its flanks compressed.[25]

Dismantling the apparatus, and examining the retort, Lavoisier found that the residue was black and well reduced. There remained "many molecules" of charcoal. The material weighed 5 ounces, 7 gros, 66 grains. A few days later the weight had increased to 6 ounces, 0 gros, 42 grains (a difference of 48 grains).[26]

A comparison of this long experiment with the similar one he had begun on May 23 reveals dramatically how much Lavoisier had learned in two months. He not only controlled almost perfectly an operation that had nearly got out of hand the first time, but designed each stage in advance more effectively. Moreover, his purpose in performing the operation was now clear. He aimed to match the weight of the minium he put into the retort with that of the air and the lead produced by the reduction.

Nevertheless, his balance sheet did not work out. The weight loss of the minium was 6 gros, 6 grains. If he assumed that the air disengaged was equal in density to that of the atmosphere, the 560 cubic inches of air disengaged in the reduction weighed only 3 gros, 41 grains. Almost half the weight lost in the process was unaccounted for. Even if he assumed that the "elastic fluid of metallic reductions, like that of effervescences, is heavier than the air of the atmosphere," and used the figure for its density that he had calculated on that assumption from his experiments on calcareous earth (see chap. 8, pp. 90–91), the air released "still weighs only 4 gros, 34 grains, and there remains, at least, a *deficit* of 1 gros, 44 grains."[27]

From his experiment of July 22 on the production of water in the reduction, Lavoisier knew already that he could not account for the discrepancy with the 24 grains collected there. What else was there to do? The problem of the density of the elastic fluid had now confronted him often enough to persuade him that it was necessary to determine experimentally the relative weights of all the "different elastic fluids." He had set out to do just that in the early weeks of his investigation, but had dropped that plan when the workers failed to complete the apparatus that he had designed for the purpose. For one reason or another, he still did not possess the apparatus he believed necessary for such a study.[28] He had no time to wait for that to happen, because he was bent on making the results of this experiment public on the very day that he completed it.

With hindsight it might appear, at first glance, that Lavoisier had missed a chance to solve his immediate problem which would not have escaped him at a later stage of his career. He had weighed the lime water in the four bottles before the operation. Had he weighed them again afterward, he might have determined the weight of fixed air disengaged directly, without requiring knowledge of its aeriform density. That solution was not open to him, however, because he identified the elastic fluid contained in the minium with all of the air disengaged, not just with the portion absorbed as fixed air. The new tests of the effects of the portion unabsorbed resolved none of his difficulties about the nature of that substance. It extinguished a candle "as fixed air would have," and it killed animals, but not as instantaneously as fixed air did. How could it be fixed air, if it evaded the absorption in lime water which constituted the most characteristic test for fixed air? Yet how could it not be fixed air, if a portion of it was fixed air, and Lavoisier also clung stubbornly to his belief that there was a single elastic fluid transmitted through all of the chemical operations that he had examined? Even as his prowess in the management of a bold new style of experimentation matured, the fissures in the theoretical structure he sought to build remained gaping.

It is easy to imagine Lavoisier rushing from his laboratory to the meeting of the Academy on July 31, the results of experiments just completed hastily written up, ready for delivery in this last of his three reports to his peers of his experimental progress. Whether that picture is real or imaginary, there is good circumstantial evidence that the experiments he had worked so hard to complete in the days just before this meeting provided the showpieces for his memoir.[29]

The main topic of Lavoisier's third memoir was "The existence of an elastic fluid fixed in metallic calces." He gave a logical, but historically misleading, account of how he had engaged himself in the topic. It was, he maintained, the experiments on metallic precipitations discussed in his preceding talk that had led him "to begin to suspect that the air of the atmosphere, or some elastic fluid contained in the air was capable . . . of being fixed in, of combining with, the metals."[30] Actually, as we have seen (in chap. 8, pp. 95–97), the experiments to which he referred were among the last things that he had done before presenting his second paper, only two weeks before. The experiments which he would present as the outcome of his interest in this question he had taken up nine months earlier than that. As best as can be ascertained from surviving documents, it was the weight gain he had observed when burning phosphorus and sulfur in October 1772, that had first persuaded him that "the augmentation of weight in metallic calces" may also be caused by the air fixed in them.[31] And the first experiment that he carried out to confirm this conjecture was the primitive version of the reduction of minium whose hard-won more advanced version he was about to report as his most recent achievement. Scientific investigations often move in circular patterns, and scientists who look back on the origins of their advances often, like Lavoisier, replace such roundabout pathways with logically reconstructed linear ones.

Lavoisier next adduced, as commonly known phenomena, several observations that gave his initial "conjecture" a "great degree of probability." They were, that metals cannot be calcined in closed vessels deprived of air, that metals whose surfaces are most exposed to the air are calcined most rapidly, and that it had been observed "by all metallurgists" that "there is an effervescence at the moment when a metallic . . . calx passes to the state of a metal." Such conjectures, he declared, can only be confirmed or destroyed by experiment. He passed over the early stages of his experimental quest in a succinct passage: "Consequently I made, in succession, different attempts, of which a great number were not fortunate, and the details of which I shall spare . . . [the audience], until finally I was able to establish the truths that follow."[32]

After we have followed Lavoisier through the multitude of failed or partially successful experiments, the changes of course, the surprises and setbacks that had befallen him before he arrived at the experiments that he was ready to present as the guarantors of truth, this covert allusion to those vicissitudes may seem more cavalier to us than to his listeners. His disinclination to describe them, however, should not be taken as an effort to conceal his missteps, nor to hold back evidence unfavorable to his case. Such references are common in the scientific literature. Scientists regularly skip over the torturous trail they have followed, when they present their views from the vantage point they have finally reached, because they do not regard the twists and turns along the way as pertinent to the conclusions they now wish to defend.

After his introductory remarks, Lavoisier launched directly into the description of his experiments on the reduction of minium. He began with the one he had carried out with the burning lens on March 30–31 (see chap. 2, pp. 25–26). "Although that first experiment was decisive enough," he claimed, it had left him with some anxiety due to the small quantity of material on which he could operate, and the large space inside the jar relative to the quantity of air disengaged. Consequently he had been "obliged to have recourse" to the large-scale Hales apparatus and iron retort that we have seen him using in the three subsequent reduction experiments. Of these, he described only the latest one that he had just completed. After giving the details of the construction of the iron retort and the arrangement of the apparatus, he summarized his procedure, reported the total volume of air released and the weight lost by the minium pretty much as he had recorded them in his notebook, and discussed the problem posed by the deficit that has already been mentioned above. He claimed to have "repeated this experiment a great many times, and the circumstances have always been very exactly the same."[33] Knowing that he had repeated the experiment only three times, and that the results had been problematic in the first two, we can see here an early tendency to create, by such vague generalizations, an exaggerated impression of the number of experiments that he had performed, without actually makng overtly false claims. That mildly deceptive practice became habitual with him.

His discussion of the difference between the weight lost by the minium and that of the elastic fluid produced led Lavoisier to the reflection that the "few drops of phlegm that I have constantly found in the recipient . . . caused me to suspect that, independently of the fixed elastic fluid, there existed in the calx a portion of water that was separated during the reduction, and that it was probably the cause of the loss of weight that I observed." Consequently, he related, he repeated the experiment with an ordinary distillation receiver to capture the water. The water received was only 24 grains, far too little to cover the deficit. Lavoisier tried to rationalize the result by suggesting that there might have been more water carried away by the elastic fluid, but he came close to acknowledging that the experiment had not confirmed his hypothesis.[34]

Recognizing that the experiment which he portrayed as following the reduction experiment had actually preceded it, we can identify another subtle way in which the rational reconstruction of a scientific investigation can make it appear more methodical than the actual pathway followed by the investigator had been. An experiment whose outcome he already knew when he encountered an anomaly, he presented instead as an experiment devised subsequently to test a hypothesis that he posed in response to that anomaly.

The last experiment concerned with the reduction of minium that Lavoisier presented was the separation of the lead from the charcoal in the residue left after the reduction in the experiment of July 22 in which he had collected the water given off into a distillation receiver. He slightly misrepresented this experiment, too, by identifying the residue as the one left after the full-scale reduction experiment; but, as the quantities of minium and charcoal employed were the same in both experiments, the substitution did not detract from the validity of the result. From this result he extracted an important further consequence. From the quantity of charcoal remaining in the residue he calculated that the total consumed in the reduction could not have been more than 1 gros, 18 grains. The elastic fluid disengaged, even making the minimum assumption that its density equaled that of ordinary air, weighed at least 3 1/2 gros. "It follows that it is necessarily at the expense of the minium, that the greater part of the elastic fluid was furnished."[35] The reason that this inference was critical for Lavoisier was that the necessity for charcoal to carry out reductions of metallic calces prevented a simple equation between the composition of the calx and the products of its decomposition. Just as the Stahlians regarded the charcoal as the source of the phlogiston that made metals metallic, Lavoisier had to treat the charcoal as a potential source of the elastic fluid disengaged. As his reasoning acknowledged, he could not rule out that source. The best he could do was to prove that it could not have been the sole, or the major source. That this problem was on his mind suggests that his abortive reduction of mercury calx without charcoal was a failed attempt to solve the problem in a more fundamental way.

In his reflections on these experiments, Lavoisier confronted the theoretical issues raised by the uncertainty over the respective contributions of the charcoal and the minium to the products of the operation:

> It appears proven, according to these experiments, that it is not the charcoal alone that produces the disengagement of elastic fluid. . . . Neither is it the minium alone, because, according to the experiments of M. Hales it gives very little air. The greater part of the elastic fluid disengaged results, therefore, from the reunion of the powdered charcoal with the minium. This last observation leads us gradually to very impor-

> tant reflections about the use of charcoal . . . in metallic reductions. Does it serve, as the disciples of Stahl think, to render to the metal the phlogiston that it has lost? Or do these materials enter rather into the composition of the elastic fluid itself? It seems to me that the current state of our knowledge does not yet permit us to pronounce on that question.

Certain experiments, which were not yet complete enough to make public, Lavoisier added, led him to conjecture that every elastic fluid is a combination of a solid or a fluid with an inflammable principle. According to that hypothesis, the substance fixed in a metallic calx would not be the elastic fluid itself, but "the fixed part of that fluid deprived of its inflammable principle." The role of the charcoal in reductions would then be to render to the fixed elastic fluid the "phlogiston, or matter of fire" that restores to it its elasticity.[36]

Although this idea might appear "remote from that of Stahl," Lavoisier believed that it "may not be incompatible with it." It was possible that the addition of charcoal both rendered to the metal the inflammable principle that it had lost, and to the fixed elastic fluid the principle of its elasticity. These matters were, however, still so delicate, so difficult, and so obscure, that it was necessary to be very circumspect. Time and experiments alone could "fix our opinions."[37]

These are the compromise statements that Lavoisier looked back on four years later as indications that he had not yet had sufficient confidence in his own views to challenge the prevailing doctrine. Time and circumstances, however, had by then so changed his perspective that he could no longer recognize this as his real position in the summer of 1773. Here he expressed publicly for the first time the same conception of the role of fire matter in the elastic state of matter that he had formulated in greater detail in his unpublished manuscript on the three states of matter in April. Here, as there, he merged his concept of a matter of fire with a modified conception of phlogiston. Only the suggestion that charcoal might contribute phlogiston both to the metal and to the elastic fluid can be construed as a tactical accommodation with Stahlian orthodoxy. There is no reason to doubt, however, that Lavoisier believed what he said.

If his position was realistic, Lavoisier's conciliatory tone still appears remote from the revolutionary spirit that had animated him in February, that had led him in the first draft of his Easter memoir to describe his theory as destructive of that of Stahl, and to predict in the final version an epoch of total revolution in chemistry. One possibility is that he had become cautious about challenging doctrines to which "all modern chemists," including his elders in the Academy, adhered. We need not resort to an explanation of timidity, however, because there was enough in his own laboratory experience over the intervening months to account for his change of mood. When he seemed ready to throw down the gauntlet in April, he believed he had a theory so powerful and so simple that it would sweep all before it. But he did not yet appreciate how hard it would be to support his theory with decisive evidence. By July that theory was already in disarray. The phenomena he had so readily attributed to fixed air as defined by Joseph Black he could now only ascribe to nothing more definite than "the air of the atmosphere, or some elastic fluid contained in the air."[38] In compensation for the loss of theoretical coherence he had suffered, he had acquired, by painstaking effort, an experimental method of steadily increasing scope and power. He had learned also, however, that the problems he hoped to solve with this method would not yield as quickly as he had once imagined. He needed to, and could afford to be, more patient than he had been just a few months earlier. In April he had announced that he could no longer defer making his theory public, for fear of losing his discovery to someone else. In July he could be content to let time and experiments "fix our opinions."

His more reticent theoretical position did not mean that Lavoisier had become less ambitious to establish, with his new venture, a prominent scientific reputation. Repeatedly, in the papers he delivered to the Academy during July, he found ways to extol his devotion to experimentation. After he had finished his presentations, an opportunity arose to publicize his research achievements more widely. Trudaine de Montigny, on whom he had already come to depend for support and guidance, suggested that, because the experiments he had conducted since beginning his research program in February were too numerous to fit into the ordinary form of short memoirs, he should publish them in a separate treatise.[39]

Lavoisier accepted this suggestion with alacrity. Sometime no later than the day after his third presentation of his results to the Academy, he informed Trudaine that he planned to collect his research into a book, and asked his approval to dedicate the book to him. On August 2, Trudaine wrote Lavoisier that "I am delighted to see that you intend to publish on fixed air. I can only be very flattered that you wish to speak of me. But I beg you urgently not to make an eloge, for which your friendship for me would surely misguide you."[40]

Trudaine was not merely a passive admirer of Lavoisier's achievements. Intimately familiar with the work itself, he acted as an influential guide and critic. It may well be that he even examined Lavoisier's laboratory notebook during the last week in July. In his letter of August 2, he outlined further experiments that he believed Lavoisier should do:

> I request you urgently, Monsieur, to do an experiment that appears to me would throw great light on the theory that one could draw from all that has been written on the different species of air. You have proved to us through very certain experiments that metals absorb a large quantity of air in their calcination. I would like to know, first, if it would be possible to test whether they are capable of calcination under the recipient of a pneumatic machine. 2) If one causes them to be calcined in a quantity of air that is insufficient to saturate the metal. . . to see if the totality of the air will be absorbed. 3) finally, it will be very essential to be able to procure the residual air left over from that absorbed by the calcination of metals and try whether it can serve for the respiration of animals, and if candles are extinguished in it. In the latter case, which I suspect will occur, it would be essential to measure by some means the length of the flame of the candle in common air and in that air, and the length of time that the flame lasts. I believe that all these experiments would throw a great light on the theory of fixed air.[41]

Trudaine's recommendations were cogent. All were directed at the central ambiguity in Lavoisier's ideas about calcination (and could as well have been directed at his parallel views on other forms of combustion). Lavoisier's inability to identify the residual air, or to decide whether atmospheric air itself, or only some elastic fluid contained in the air, was fixed in metals in calcinations were just the problems that Trudaine's suggested experiments were intended to clarify.

Lavoisier undoubtedly understood the force of his friend's advice, but he was unwilling to delay the completion of the projected book until such experiments could be undertaken. Perhaps, as Jean-Pierre Poirier has recently suggested, his anxiety to publish quickly was based on continued concern that other European scientists working on fixed air might forestall him.[42] In any case, he carried out only a few more experiments during the next week, to fill out gaps in series already performed, while he rushed to complete his manuscript.

Lavoisier divided his treatise into two parts. Part 1 consisted of the historical summary of previous experiments and opinions about "Elastic emanations disengaged from bodies during combustion, fermentation, and effervescences" that he had delivered in three lectures to the Academy in April and May. Part 2 presented his own "New Researches" on the existence of an elastic fluid and the phenomena resulting from its disengagement or fixation. He converted the three memoirs delivered in July to the first four chapters of this part, then added further chapters on the experiments he had not covered there. With the exception of some complete failures and first tries on which he had later been able to improve, Lavoisier managed to incorporate into his treatise almost every experiment he had performed since the beginning of his research program in February.[43]

During what must have been an intense effort, mostly at his writing desk, Lavoisier put together his treatise during the first week of August. For the chapter on the combustion of phosphorus he went back to his laboratory for a final effort to weigh the acid free from moisture. This time he placed the same quantity of phosphorus used in the previous experiments (8 grains) into a large crystal bottle closed with a "very good cork stopper." He weighed the bottle with its contents and stopper very carefully, assuring himself that the total was, within half a grain, exactly 15 ounces. He then placed the bottle under a glass jar, taking care to remove the stopper just an "instant" before he plunged the inverted jar into mercury in the pneumatic basin. Inflaming the phosphorus as usual with the burning glass, he observed, as before, the formation of white vapors which attached themselves as flakes to the glass surfaces. About two-thirds formed in the bottle, the rest on the inverted jar. The mercury rose 14 1/2 lines, representing an absorption of 16 3/4 cubic inches of air. Removing the inverted jar after the combustion was complete, he was able to stopper the bottle again "in 4 seconds." Convinced that this was so short a time that the dry air in the bottle could not have been mixed with humid air from outside, he weighed the bottle again, confident that the 7 grains of weight gained in the formation of these flakes was due to air alone. The experiment would have been a complete success, if only all the flakes had collected inside the bottle. The one-third that collected in the outside jar, he estimated, would have gained proportionately 3 to 4 grains if they had been in the dry air of the bottle. The total would, therefore, have been 10 to 11 grains. "That is," he inferred, "more than the weight of the air absorbed, which proves there was an absorption of water that contributed."[44] Apparently Lavoisier figured the weight of the air by assuming it equal in density to ordinary air, despite the evidence from his other experiments that it might be heavier. Foiled once again in his attempts to prevent any absorption of water after the combustion was completed, he had nevertheless capitalized on his partial success to draw the meaningful conclusion that some water is absorbed in the combustion process itself.

On August 7, Lavoisier performed one last experiment on the combustion of phosphorus, placing the material in the same large bottle that he had used in the experiment of July 20 to try to weigh the product before it had had time to absorb moisture. This time, however, his attention was directed less to that problem than to the question of what would be the effect of burning more phosphorus? In place of the 8 grains that he had used each time until now, he placed 17 2/3 grains in the capsule in the bottle. The total weight of phosphorus and bottle before the operation was 6 ounces, 4 gros, 4 grains. When he set the phosphorus on fire with the burning lens, white clouds arose as usual, and air was absorbed up to "a point, after which the phosphorus was extinguished." In all the previous operations the phosphorus had burned completely. Lifting the jar off the mercury, he quickly reweighed the bottle, and found it to be 6 ounces, 4 gros, 10 1/2 grains. The gain had been about 5 to 6 grains. To verify that unburned phosphorus remained in the capsule, he uncovered it under some of the flakes of acid, and relit it. It burned "for a long time."[45]

"It is clear," Lavoisier wrote, "that in this experiment no more phosphorus was burned than in the preceding one, even a little bit less."[46] That was a dramatic result. Had he expected it, or was it a surprise? There is sufficient similarity between the experiment and Trudaine's advice to him five days earlier to see if one could calcine a metal in a quantity of air insufficient to saturate it, to surmise that Lavoisier *might* have transferred the suggestion from calcination to combustion. He need not, however, have needed that influence to raise the question for himself. Having already expressed uncertainty over whether the elastic fluid whose identity he sought was the air itself, or only something contained in the air, he could have arrived at the inference that an experiment designed to test the limits of combustion in a given air space might help to answer that question. On the other hand, the chapter "On the combustion of phosphorus and the formation of its acid," in which he included this experiment, gives little reason to think that he had framed the question clearly. After discussing the fact that the quantity of phosphorus burned in this case was no greater than in the other experiments in which he had used smaller amounts of it, he concluded merely that "these experiments seem to lead [one] to think that the air of the atmosphere, or some other elastic fluid contained in the air, combines during combustion with the vapors of phosphorus."[47]

By reversing the order in which he discussed two of the other experiments he had performed on phosphorus, Lavoisier again subtly enhanced the logical sequence of his quest to "determine, with as much precision as this type of experiment allows, the augmentation of weight of the acidic vapor of the phosphorus that burns."[48] By placing his latest effort to prevent absorption by putting the phosphorus into a stoppered bottle just before his account of the experiment of July 21 in which he had placed a capsule of water beside the capsule of phosphorus under the inverted jar (see above, pp. 106–7), Lavoisier made the chronologically earlier experiment appear to be the outcome of reasoning stimulated by what was actually the later one. From the fact that the phosphorus had gained 10–12 grains in the stoppered bottle experiment, Lavoisier no longer drew the conclusion that he had in his laboratory record, that "absorption of water" must have contributed, but inferred instead that "the elastic fluid absorbed weighs around 2/3 of a grain per cubic foot, that is, almost a quarter more than the air that we respire." But, he asked himself, why might it not be water vapor itself, held in solution by the elastic fluid, which makes it "the heaviest part of the air"? In that case, water must be necessary for the combustion, and if it is removed, the combustion cannot take place. This idea seemed so seductively probable to him, Lavoisier related, that he hurried to subject it to the test of experiment. "Here is the reasoning that I went through." 1) If water were supplied as vapor to replace that absorbed in the combustion, the combustion should be prolonged. 2) There should be no further decrease in the volume of the air as the phosphorus continues to burn. "These reflections," he asserted, "led me to the following experiment."[49]

The experiment was "to burn phosphorus under a jar plunged into mercury, while supplying in the same jar an atmosphere of water reduced to vapor." The result did not differ in any way from all others made over mercury except for the fact that the acid collected in drops rather than in concrete flakes. He concluded that the diminution of volume observed in the combustion could not have derived from the absorption of water. He did not deny that the gain in weight beyond that accountable to the air absorbed came from the water, but the "greater part of the augmentation of weight of phosphoric acid is due to a substance other than water."[50] Lavoisier's reasoning was impeccable in every respect except that it could not have been this sequence of reasoning that led to an experiment he had already performed.

By August 7 Lavoisier had stitched together his volume. At the meeting of the Academy that day, he asked that commissioners be appointed "to examine a treatise on fixed air that he wished to publish." This duty was assigned to Le Roy and the pharmacist Louis-Claude Cadet de Gassicourt.[51]

* CHAPTER TEN *

Mopping-Up Operations

THE COMMISSIONERS assigned to examine Lavoisier's treatise decided to verify his results by asking him to repeat in their presence the most important of his experiments. For this purpose Trudaine de Montigny and Pierre Joseph Macquer were added to the commission. Witnessing experiments they were called upon to evaluate was a common procedure in the Academy, but if Lavoisier were as anxious to make his results public as his rush to finish his treatise suggests, he may not have welcomed the delay.

Trudaine, Le Roy, Macquer, and Cadet assembled in Lavoisier's laboratory on Saturday, September 25, for the "Verification of the experiments of M. Lavoisier on the fixation of air in bodies and the elastic fluid that is disengaged in various circumstances." Macquer wrote out the procedures and results in Lavoisier's current laboratory notebook. Following the general order in which he had arranged the topics in his treatise, they carried out representative experiments of each type, beginning with the dissolution of chalk, slaked lime, and soda in nitrous acid. They did two experiments to measure the quantity of elastic fluid contained in lime, then the quantity disengaged from soda by nitrous acid. They changed concrete volatile alkali to the caustic alkali by adding slaked lime, and measured the density of several of the solutions with the areometer. They repeated the combustion of phosphorus, using Lavoisier's maneuvers with the stoppered bottle to try to reduce the absorption of water.[1]

Not all of the experiments Lavoisier carried out with his examiners were mere duplications of his previous work. Instead of doing over the paired experiments on the dissolution of soda in nitrous acid through which he had measured the fixed air disengaged directly and indirectly (see chap. 8, pp. 87–89), they performed comparable experiments on volatile alkali, thus filling in places within the general scope of his research plan that he had not previously had time to occupy. Another novel twist was to precipitate lime water with the vapors of burning charcoal.[2]

The most ambitious experiment the Commissioners verified was the reduction of minium. The apparatus and procedures they used were those that Lavoisier had devised over the course of the three experiments he had carried out with the iron retort and large inverted vessel. Macquer was impressed with the operation itself; "*Immediately* after one started the fire in the furnace," he recorded, "the water descended, almost in proportion to the volume of air contained in the retort. Soon the base of the retort began to redden, and there was a considerable descent of the water, rapid enough to be apparent to the view."[3]

The Commissioners were particularly interested in the properties of the air disengaged. To this end they modified Lavoisier's earlier procedure, passing the air first directly into a large collection bottle. When they plunged a lighted candle into this air, it was extinguished. They repeated the maneuver seven or eight times, noticing how the flame behaved when they lowered the candle slowly into a bottle containing the air.[4]

Next they passed some of this air through two lime water bottles into another collection bottle. A lighted candle immersed in the air now was extinguished "much less promptly," but they observed that when the lime had been precipitated and no longer absorbed the elastic fluid, the air which continued to pass through the bottles "became more effective in extinguishing the flame." To extend the comparison, they introduced one sparrow into a bottle full of air that had not been passed through the lime water, and another into air that had been. The first sparrow collapsed "in an instant as if dead." Retrieving it and placing it in a window, however, they revived it, and it flew away. The sparrow placed in the air passed through the two lime water bottles endured for a longer time, but would also have died if they had not removed it. They placed a third, stronger sparrow in a larger bottle of the air not passed through lime water. In twenty seconds it was dead. After describing these extended tests, Macquer recorded the disassembly of the apparatus, measurement of the weight of material left in the retort, and calculation of the total quantity of air disengaged, as routine procedures.[5]

The clarity with which these comparative tests were performed, and the distinctions noted, contrast vividly with the confusing, indecisive tests to which Lavoisier himself had subjected air disengaged in the reduction of minium in the experiments he had conducted on his own. In the light of Trudaine's letter of early August urging him to carry out similar tests on the residual air left from the calcination of metals, and

Trudaine's presence during these tests, it is not difficult to see his hand in this significant improvement of Lavoisier's experiment. The conviction that there was only one elastic fluid involved in all the operations that he was studying was for Lavoisier a strong mental obstacle to the recognition of the differences that these new tests revealed between fixed air and whatever else was released in the reduction of minium. If Trudaine really was the person who induced Lavoisier to make these changes in his experimental procedures, then there is real substance in the tribute Lavoisier offered his older friend and adviser in the dedication of the volume which emerged from these trials: "It is you . . .who have guided me more than once in the choice of experiments, who often clarified their consequences for me; finally, who desired that the majority of them be performed or repeated before your eyes."[6]

The Commissioners had now been working steadily for several days with Lavoisier. On Tuesday, the 29th of September, they made with him two more significant efforts to extend the scope of the experiments they had only been charged to verify. To pursue the question whether the charcoal might contribute some of the air disengaged in the reduction of minium, they heated the same quantity of powdered charcoal that he had used for the reduction, but alone in the apparatus. The water level descended, and the air produced precipitated lime water, but "more slowly than that from the reduction of minium." After the operation the charcoal was found unchanged in appearance, but had lost 24 grains.[7] The result was probably regarded by the group as indecisive, but suggestive enough to invite further experiments.

One more experiment, so simple that it may have been tried on the spur of the moment, Macquer recorded under the heading "Experiment to reexamine":

> Air of respiration—introduced by a glass tube into lime water. Precipitates very promptly.
>
> Atmospheric air introduced into lime water by a bellows did not trouble or precipitate the lime water.[8]

These tests may have made an impression on Lavoisier out of proportion to their brevity. When he had sketched out his experimental program in February, respiration was one of the "operations that fix air" with which he had wanted to begin. Diverted from that plan when the craftsmen failed to produce the apparatus he had designed for that purpose in March, he had subsequently been so preoccupied with his other problems that he seemed to have lost sight of that part of his plan. Whether he, or one of his colleagues suggested this little comparison, the strong contrast it revealed between ordinary air and air which had been breathed probably helped to revive Lavoisier's interest in this subject.[9]

Lavoisier and the commission conducted one more experiment on the combustion of phosphorus. This time they duplicated his procedure for burning the substance in air saturated with water vapor, by placing a capsule of water under the jar together with the capsule containing the phosphorus. Macquer described the rise of the mercury and the collection of the phosphoric acid in droplets, without further comment. The Commissioners were either satisfied that they had verified Lavoisier's procedure, or at least, came to appreciate how difficult it was to determine accurately the nature of and the exact quantity of the weight gained in the combustion. They did not attempt to weigh the product.[10] With this they ended their oversight of the experimental operations on which their young colleague had based the claims made in his treatise.

The benefit to Lavoisier from these experiments carried out in collaboration with his senior colleagues extended beyond its role in obtaining the endorsement of the Academy for his publication, or the contributions Trudaine and the others may have made to the extension and clarification of his results. The direct familiarity they acquired with his innovative quantitative experimental approach, just at that stage when it had matured into a powerful method, could not help but strengthen his position for the challenges to prevailing chemical doctrines that he knew lay somewhere ahead of him. Particularly strategic for him was the presence of Macquer, one of the most prominent and respected of the older chemists in France. A staunch believer in the central importance of the phlogiston theory of Stahl to chemistry,[11] Macquer would be among the most strategic figures to be won over if Lavoisier were to move from the cautious compromise views he currently espoused toward the more confrontational position with which he had flirted in the first flush of enthusiasm for his "new theory." Respect for Lavoisier's methods in the laboratory would stand him in good stead, if he should come to differences with Macquer over the theoretical interpretation of his results. There is good evidence that he did attain not only Macquer's respect, but his strong admiration for these experiments. Three months later, as the publication that he was examining was about to go into print, Macquer corresponded with the distinguished Swedish chemist Torbern Bergman about a memoir

Bergman had published on fixed air. "We are beginning to work very much here [in Paris] on that curious and important matter," Macquer wrote. "There has just appeared a work on that subject by M. Lavoisier, one of our Academicians, which is very well made and which I believe will satisfy you, if it reaches you."[12]

Meanwhile Lavoisier had also continued, on his own, to refine and extend the experiments pertinent to his treatise. In August he had, among other things, examined further the passage of air from volatile alkali into lime and from chalk into lime, the dissolution of lime in water, and the precipitation of lime water by the air disengaged in the reduction of minium. He redetermined the relation between the weight of lead and of minium by burning the residue resulting from one of his reduction experiments in a ladle.[13] In September he tried some more precipitations of metallic solutions.[14]

After the Commissioners had verified his experiments, Lavoisier began to direct some of his attention to other problems, including participation with his colleagues in further experiments on diamonds with the great burning lens in the *Jardin de l'infant*.[15] Several times in October, however, he returned to the subject of his impending publication to repeat experiments he thought could still be improved, or to pursue questions raised by the experiments performed with the Commissioners.

On October 19, Lavoisier pushed further the effort begun with the Commissioners to measure the elastic fluid that charcoal might produce by itself. To enable him to apply as much heat as possible, he substituted a gun barrel for the iron retort. So that it would fit into his apparatus for collecting the air, he bent the barrel into a curve. Putting 3 gros of charcoal inside the barrel, he connected it up and kindled a very strong fire around it. At the beginning, through "some unknown circumstance," the water level fell a little, leading him to suspect there was a small hole in the barrel. He heated the barrel until it became reddish white. At first there was a rapid and considerable disengagement of air, but it soon ceased. He continued heating for an hour and a half, but no more air was released. By then he decided that there was no sensible error due to the possible leak, and that anyway, if there were, it would be "against me." The total volume of air produced was 132 cubic inches. The next morning he found 2 gros, 15 grains of charcoal in the gun barrel.[16]

"At once" Lavoisier mixed this charcoal with 6 ounces of minium and put them back in the gun barrel. The volume of material was too large to fit in the lower end of the barrel. As soon as he began heating, air was rapidly disengaged. It soon stopped, but by shaking the barrel so that more of the mixture fell into the heated end, he produced another burst of air. By repeating this maneuver three or four times, he obtained 448 cubic inches of air. This was sufficiently greater than the 132 cubic inches produced by the charcoal alone to confirm the view he had already made public, that the greater part of the elastic fluid derived from the minium. He wondered if the air disengaged in the first of these experiments had come from some iron rust that may have been in the barrel, even though it was new and he had cleaned it thoroughly.[17]

Afterward, Lavoisier found "about 1 ounce, 2 gros" of minium still in the barrel. If a proportional amount of air had been disengaged from 6 ounces, the result would have been 566 cubic inches. With satisfaction Lavoisier squeezed into his record of the experiment, "that would come again very exactly to the result that I have given in my memoir."[18]

After spending several days on experiments connected with the diamond project, Lavoisier came back, finally, to a critical aspect of his research on the fixation of elastic fluids that he had neglected ever since his discouraging early attempts in March and April. That was, to demonstrate that metals gain weight and absorb air in calcination. He resorted again to the burning lens to heat the metal, but now inverted the vessel in which the operation was to take place in mercury in place of the water covered with oil on which he had relied before learning about the advantages of mercury. To reduce the amount of mercury necessary, he substituted, for the jar he had used then, an inverted cucurbite, a traditional vessel normally used for distillations, whose narrow neck would permit him to place the mercury reservoir in a small basin.

His modified arrangement still permitted Lavoisier to calcine only a small quantity of metal. In his first attempt, on October 26, he put only 1 gros of lead in a very small porcelain capsule. When he focused the burning glass on it, the surface calcined "promptly." Soon the calx seemed to melt into the surface of the capsule. Yellow and grey vapors arose and attached themselves to the walls of the cucurbit. After about an hour he had to stop the operation, because the sun became hidden. The mercury had risen two lines. Perhaps responding to Trudaine's earlier recommendation, he put a candle into the residual air. It went out. Testing the air with lime water, he found that the latter was "not precipitated. Nevertheless, after quite long agitation it became a little cloudy. But might this circumstance derive from fixed air produced by the candle that had been introduced?" It is hard to guess what Lavoisier thought at this point the residual air might be.[19]

The next day Lavoisier repeated the calcination attempt. To avoid the problem he had encountered when the calx seemed to melt into the capsule, he substituted for the latter a "strong earthen plate" with a flat bottom to increase the surface that he could expose to the burning lens. This time he employed 3 gros of lead, probably to make the new experiment comparable to the one he had performed in March. Because the sun was weak, he tried to keep the lead not directly in the focus of its rays, but a little above it. He had taken his lead from the center of a large mass, to avoid impurities, but as soon as the material melted, a pellicle formed on its surface. As the calcination proceeded, the pellicle turned yellow. Cracks formed on the surface, from which small globules oozed. They were metallic in appearance, but he thought they were massicot. Further changes took place in the appearance of the material, but after about ten minutes the calcination was arrested. Some of the material seemed to evaporate, a process that he tried to stop by keeping it out of the focus of the lens. He continued the operation for an hour and forty-five minutes, but because the sun came in and out of the clouds, he estimated the exposure at about an hour.[20]

After the vessel had cooled, the mercury stood at 2 1/2 lines above its initial mark, from which Lavoisier evaluated the absorption of air as 3 3/4 cubic inches. He took the density of the air to be "2/3 of a grain per cubic inch," which would give 2 1/2 grains for the weight of the absorbed air (a result which he did not write down). The weight of the lead had been augmented by "1 grain, at least." The difference between these two figures he guessed to be due to evaporation, which he estimated to have amounted to at least one grain. (Afterward, he corrected the gain in weight of the lead to "1 grain and three-quarters, at least," and changed the loss from evaporation correspondingly to 3/4 grain.)[21]

Turning the cucurbit over, he plunged a candle into it. It burned at first, but gradually languished and went out in less than a minute. Lime water did not precipitate, and scarcely became cloudy. Whatever the residual air was, it was not fixed air.[22] His use of the number 2/3 for the density of the air absorbed indicates, however, that he was now assuming that his elastic fluid was heavier than ordinary air.

On the same day, Lavoisier repeated the experiments on charcoal and the reduction of minium with a new, unused gun barrel. The results persuaded him that the air released from the charcoal could not have come from iron rust.[23]

The relative success of his latest effort to calcine lead must have been very satisfying to Lavoisier, because it reinforced what had been, until then, the weakest link in the chain of his experiments on the elastic fluid fixed in metals. Moreover, this result transformed his earlier calcination attempts from failed experiments into evidence complementary to his new experiment. When he wrote, or rewrote, the chapter on "The combination of the elastic fluid with metallic substances by calcination," he began with accounts of the calcination of lead, tin, and the alloy of lead and tin that he had performed in the spring, then followed them with his most recent experiment. He had no reason to conceal the fact that little or no calcination had taken place in the earlier experiments, because the fact that they had neither gained appreciable weight nor absorbed definite quantities of air now fit the general proposition that

> in the measure that the calcination takes place, there is a diminution in the volume of the air, and this diminution is almost proportional to the augmentation of the weight of the metal.[24]

He really had only two points along the scale of this "measure": the result of his latest experiment which gave him a proportion of "2/3 grain per each cubic inch of air,"[25] and a base point in which there was no weight gain or absorption. So convinced was he of the validity of his interpretation of the process, however, that he needed no more.

The fact that the calcination in his latest experiment had ceased while some of the lead still remained unchanged became the principal foundation for another general conclusion:

> That this calcination has limits; that is, that when a certain portion of metal has been reduced to calx in a given quantity of air, it is no longer possible to carry the calcination further in the same air.

That conclusion, supported also by a long-term experiment in which iron immersed in water in a closed space rusted for about two months and then stopped, led Lavoisier in turn to a proposition that appeared to resolve his indecision about whether the elastic fluid he was tracking was the air itself or something contained within it:

> That several circumstances would seem to lead one to believe that all of the air that we respire is not suitable to enter into combination with metallic calces; but that there exists in the atmosphere a particular elastic fluid that is found mixed with the air, and that it is at the moment at which the quantity of this fluid contained under the jar is exhausted that the calcination can no longer take place. The experiments [on the combustion of phosphorus] that I shall report in chapter IX give some degree of probability to this opinion.[26]

Taken by itself, this statement suggests that his experiment carried out in October, nearly the last one performed to complete the evidence he would incorporate into the final revisions of his treatise, had led him to a critical advance in his theoretical structure. But did he really believe this conclusion himself? In the chapter on the combustion of phosphorus to which he referred, which he had probably written earlier, he let stand the statement that the experiments "seem to lead [one] to think that the air of the atmosphere, or some other elastic fluid contained in the air, combines during the combustion with phosphorus."[27] If his final experiment with lead had carried him in the direction of a decision between these alternatives, he had evidently not embraced it with sufficient conviction to revise what he had already written.

After these experiments, Lavoisier was apparently content that he had done all he could to strengthen the experimental evidence supporting his treatise. He turned to other investigations, on Epsom salts and the affinities of phosphoric acid, that seemed to lead him toward a more classical style of chemistry. Prompted by advice from Trudaine, he interrupted these studies to return once more to a set included in the work still under examination by the Commissioners. In his notebook he commented, "M. de Trudaine having suggested to me some anxiety about the disengagement of volatile alkali from sal ammoniac by calcareous earth precipitated in caustic and noncaustic form, I have again repeated all of the experiments."[28]

The last two, brief chapters with which Lavoisier closed his treatise contained experiments that seemed not to fit in the earlier chapters. Chapter 10 gave succinct descriptions of experiments on combustion and the detonation of niter in the vacuum of a pneumatic machine. Neither phosphorus nor sulfur would burn in the evacuated recipient, and gunpowder would not ignite or detonate.[29] His cursory presentation, with no comment on the outcome, suggests that Lavoisier may have performed and included these experiments mainly to satisfy the request that Trudaine had made in August to do so.

Chapter 11 described the effects of the air in which phosphorus had been burned, on animals, and on a lighted candle. The first was extracted from the experiments performed with the Commissioners, in which a bird placed in the air seemed not to suffer, although its respiration seemed to him "more difficult than in ordinary air." He recalled that an animal of the same species placed in fixed air would have "perished at the first inspiration." A candle, on the other hand, was extinguished. The last experiment, "To mix a portion of the elastic fluid from effervescences with air in which phosphorus had burned," he had carried out, he explained, because

> I was curious, in regard to views about which I shall give an account at another time, to observe whether the mixture of one-third of the elastic fluid from effervescences would correct the air which has served for the combustion of phosphorus and restore to it the property of supporting an inflamed body.

When he tried it, however, a candle was instantly extinguished."[30]

Lavoisier's reticence about what made him curious to try this experiment leaves an intriguing gap in our understanding of his thought at this point. That one air which could not support combustion could alter another air with the same property in such a way that it afterward could support combustion suggests an interactive relation between different "airs" that seems to come from outside his search for a single elastic fluid. That he used the word "correct the air," rather than enter some "combination" with it, implies a malleability of properties at odds with the idea of stable distinct species of airs. Perhaps it is a measure of the uncertainties created by the confusing mix of similarities and differences he had noticed when he applied the simple tests available to the airs disengaged in, or left as residues in his experiments, that he entertained such conceptions.

The four Academicians ordered to examine Lavoisier's treatise, now entitled *Opuscules physiques et chymiques*, reported their evaluation to the Academy of Sciences on December 7. Because a copy of their report in Lavoisier's hand has survived, historians have assumed that Lavoisier himself drafted it for his examiners.[31] Even so, that does not mean that he orchestrated their response to his work. Ambitious and self-promoting though he was, he was not yet powerful enough to compel the assent of distinguished senior Academicians such as

Le Roy, Trudaine, Macquer, and Cadet, to his own opinion of his achievements. Although generally laudatory, the report contains a few ironic passages that establish some space between Lavoisier's aims and their assessment of his work. Of the experiments in part 2, for example, they commented that "he supposed in some way that the elastic fluid was only suspected [to exist] and undertook to demonstrate its existence and properties through a numerous sequence of experiments." Consequently "those with which he begins are not fundamentally new." They went on to justify such repetition and to praise the additions that he had made to the existing knowledge of the elastic fluid by measuring directly the quantity disengaged in the operations. Their summary did hint, however, that Lavoisier's account of his predecessors may have served to make the existence of fixed air seem more problematic than it was, so as to reserve for himself the delivery of the critical proofs in its favor.[32]

The heart of the Commissioners' evaluation was not the long abstract of the *Opuscules* that took up most of their report, but their statement that "we believe that we must assure the Academy that M. Lavoisier has repeated almost all [of the experiments] with us, and we join to this report the account that we have made of them as they were done, signed by us." They added that Lavoisier had "submitted all of his results to measurement, to calculation and the balance: a rigorous method which, fortunately for the advancement of chemistry, begins to be indispensable to the practice of that science." Having witnessed his experiments themselves, they were in a strong position to verify that he had performed them "with exactitude."[33]

The Commissioners approved of the style in which Lavoisier had presented detailed summaries of numerous experiments without extensive theoretical discussion. Some of the remarkable phenomena that he had observed gave rise

> to new and bold ideas; but M. Lavoisier, far from delivering himself too much to his conjectures, contents himself to propose them once and in a few words, with all the reserve that characterizes enlightened and judicious physicists.[34]

If the Commissioners recalled the very different style in which Lavoisier had, only eight months earlier, announced a sweeping new theory and heralded a coming revolutionary epoch, they might not only have welcomed the greater reserve he had since acquired, but wished to express their preference for this more judicious approach.

The Commissioners accepted Lavoisier's view that the "elastic fluid disengaged from the reduction of minium has exactly the same properties as that exhaled during effervescences from the combination of calcareous earths and alkalis with acids."[35] Consistent with the greater attention paid to the different properties of the air derived from metallic reductions before and after passing it through lime water when Lavoisier performed these experiments with them than before, the Commissioners commented more pointedly than he had on his observation that animals died less promptly in the one than the other. The elastic fluid derived from metallic calces, they thought, contains a portion of common air "a little greater" than that disengaged from chalk. "Finally," echoing Lavoisier's own uncertainty, they stated that

> nothing puts [us] yet in a position to decide whether the combinable part of the elastic fluid of effervescences and reductions is a substance essentially different from the air, or whether it is the air itself to which something has been added or from which something has been removed, and prudence demands the suspension of judgment on that subject.[36]

That was a somewhat clearer statement of the alternative possibilities than Lavoisier had given in his own text. More struck than he himself seemed to be by the fact that an animal placed in the air left after the combustion of phosphorus in a closed space did not perish instantly as it did in the air of effervescences, even though both airs extinguished candles, the Commissioners stated that "this remarkable circumstance indicates that there are still many important things to discover about the nature and effects of the air and the elastic fluids that one obtains in the combinations and decompositions of many substances."[37]

His examiners treated Lavoisier's work not as a finished investigation, but one still in progress. They reported that he regarded his experiments on the calcination of metals merely as a rough beginning, despite his multiple experiments—as it certainly was, even though his own presentation treated the results as decisive enough. Their recommendation to the Academy judged his work as a whole to be an auspicious beginning rather than a finished achievement. "One cannot exhort M. Lavoisier too strongly," they concluded, "to continue this series of experiments already so well begun, and we believe that the work of which we have given an account merits being printed with the approbation of the Academy."[38]

Acting now as energetically to place his work before the eyes of the public as he had done to produce it, Lavoisier arranged with three publishers, Durand on rue Galande, Didot on the quai des Augustins, and Esprit at the Palais Royal, to print the volume. One month after the Academy had approved his *Opuscules physiques et chymiques*, he was able to present to that body a copy of the published volume.[39]

In a notice inserted at the beginning of the *Opuscules*, Lavoisier explained that he had chosen this title, not because it referred to the contents of the present volume, but because he intended it to be the first of a series of volumes that would collect together the diverse investigations in physics and chemistry with which he had been occupied for "more than ten years." In future volumes he hoped to include studies he had made ranging from the use of spirit of wine in the analysis of mineral waters to "some points in optics," the height of the principal mountains in the vicinity of Paris, and the profile of the interior of the earth in the provinces of France.[40] The list suggests how broadly inclusive the meaning of the word "physics" was for Lavoisier.

The three publishers printed 1,250 copies. Trudaine received 30, and 12 went to the court.[41] Lavoisier spared no effort to get other copies into the hands of persons and institutions strategic to his reputation. He accompanied them with letters carefully composed to appeal to their vanity and, where appropriate, to the parts they had played in the development of the "truths" to which he claimed only to have added "a very small number of new" ones. When writing to the Royal Society of London, he stressed that "it is in the heart of England that the doctrine of fixed air was given birth," and made sure to refer to the "series of important experiments" that Joseph Priestley continued to pursue. He began his letter to the Royal Society of Edinburgh by congratulating them for having, among their members, "the illustrious savant who was the first to join into a corps of doctrine the phenomena of the fixation of air in bodies." It was there that a "new theory that seems to prepare a revolution in physics and in chemistry" was "almost entirely formed." To make certain that his point was not missed, he enclosed an additional copy for Black, "asking you to assure him that there exists no more zealous admirer of his talents than I."[42]

The reappearance of the theme of a revolution in physics and chemistry in his letter to the Edinburgh Society is particularly interesting, because it echoed the phrase that Lavoisier had entered in his laboratory notebook at the beginning of the investigative venture whose first stage he had now completed. At the same time, it shifts the meaning Lavoisier attributed to the phrase, and gives us a different sense of what he had meant in the first place. The statement in his laboratory notebook that he would "take up again all this work which has seemed to me made to occasion a revolution"[43] (see chap. 2, p. 15) has generally been taken as a prediction that his own discoveries, and particularly the challenge to Stahlian theory that he soon considered but then deferred, would be the source of such a revolution. But the "work" which he planned to take up was that begun by Hales, Black, and their successors, and to which he now claimed to have made a "small" addition. In his letter to the Academy of Sciences in Berlin he wrote that his book was intended to "fix the attention of Savants on an important theory which seems to open a new pathway to physicists and chemists, finally to add to truths little known some truths that I believe to be new."[44] This was not merely the conspicuous show of modesty mandated by the contemporary style of formal communication. The "important theory" was the theory of fixed air, not the "new theory" of his own that Lavoisier had prematurely announced in April 1773. The content of his *Opuscules*, his own opinion, and that of the Academicians who had examined his work, concurred in the judgment succinctly expressed here. What Lavoisier aimed to do, early in 1773, and what he felt he had managed to do by the beginning of 1774, therefore, was not to *initiate* a revolution, but to *participate* in one that he believed his predecessors had already prepared.

In his letter to the Academy of Sciences in Berlin, and in several others, Lavoisier added a plea for understanding:

> Accustomed to pursue experiments and observations requiring the hand, you will easily appreciate that it is rarely possible in physics and chemistry to arrive at demonstrations as rigorous as those in geometry, and I hope that this reflection will stimulate in you indulgent feelings in my regard. Nevertheless, by multiplying experiments, by varying them, by arriving at the same consequences by different routes, one can succeed in obtaining a degree of probability that can equal certitude. It is that route that I have been forced to follow in the work that I have the honor to offer you.[45]

This is a richly revealing passage. In a text written near the end of his life, Lavoisier complained that the courses of chemistry that he had taken in his youth were, in spite of the clarity with which his teachers had professed the subject, surprisingly "obscure." He had been accustomed, he wrote, to the rigorous reasoning of mathematics, especially the "most sublime verities of transcendent geometry." In chemistry "it was a completely different route." One began by supposing what still needed to be proven. He was presented with words whose meaning required knowledge still foreign to him, and which he could acquire only by studying "all of chemistry."[46] Historians have interpreted these remarks, made long afterward, to support the view that Lavoisier distanced himself from the chemistry of his time, importing from physics, or elsewhere, the precise methods that he did not find in the chemistry he was taught.[47] The above passage shows Lavoisier instead associating himself with a chemistry that could not be pursued with the same rigorous mode of demonstration from which sublime theorems of geometry emerged, but must be dealt with by methods appropriate to the nature of its problems. Undoubtedly, in the freshness of his first youthful exposure to chemistry, after he had finished his training in elementary mathematics, and "followed the experiments of the Abbé Nollet" that may have appeared particularly elegant to him, he was dissatisfied. By January 1774, Lavoisier had learned some hard lessons. The beautifully rigorous style of demonstration that he admired in geometry would not work in chemistry. The many setbacks that he had encountered as he tried to gather experimental evidence to support a brave new theory had taught him that he, too, must follow "another route."

* CHAPTER ELEVEN *

Conclusion

TO twentieth-century readers, the *Opuscules physiques et chymiques* has seemed paradoxical. Viewed as the first move toward the revolution Lavoisier later led to completion, it appears hesitant, mild, and unfocused. Andrew Meldrum wrote in 1930 that "the *Opuscules*, at the first reading, is a disappointing work. One expects great things of Lavoisier."[1] Its lack of the militant tones one expects of the opening attack on the bastions of an establishment have led some historians to infer that Lavoisier was keeping his true intentions secret. Maurice Daumas wrote in 1941 that "the *Opuscules* is not a work of combat. It contains no demonstration, no discovery, no conclusion." His real conclusions, according to Daumas, Lavoisier shared only with a few friends. Already he had "measured the importance of the assault that he would make," but held back his intentions so that his adversaries would not have time to reinforce their positions in advance.[2] Lavoisier himself contributed to this view by his claim in 1777 that he had been deterred from challenging Stahl in the *Opuscules* only by lack of confidence in his views. From the moment Lavoisier opened his attack on the Stahlian theory in his "Memoir on Combustion in General" in that year, he looked back on all of his investigations on the fixation and release of airs before then as the prelude to that decisive move. Historians have continued to follow his example.

If we reexamine the *Opuscules*, not from the viewpoint of what later followed from it, but of the events that produced it, its format and content become less strange. If my reconstruction of the way its chapters were put together from texts written at different times, between the spring and late fall of 1773, while Lavoisier's ongoing experimentation continued to alter his views, is valid, the absence of a coherent theoretical framework for the experimental section becomes an expected outcome. Rushing to get what he had done in print, he could not wait long enough to revise thoroughly and organize into one integrated argument the partial arguments that framed each topic.

It was not haste alone, however, that prevented Lavoisier from presenting a unified theoretical structure. The beautiful theory that fixed air is combined with metals in calcination and released in reductions had collapsed under the assault of his own efforts to provide the supporting evidence that he had lacked when he announced it, with youthful self-assurance, in April 1773. When he completed the *Opuscules* he had no finished theory to replace what he had lost. He was probably not unhappy that the fragments of a theory that he hoped eventually to complete were scattered inconspicuously in the "reflections" at the end of each chapter focused on one of the experimental problems at which he had worked so diligently during the past year.

What Lavoisier had lost theoretically since April, he had gained experimentally. In the spring he was still emulating the methods of Hales and Black, from whom he took the rudiments of a quantitative method. He did not yet know either the pitfalls into which his method could lead him, or the scope of its possibilities. By the end of that year, he had forged from these elements a style of quantitative experiment and reasoning that already distinguished him from his predecessors.

Historians have sometimes compared Lavoisier with Joseph Priestley, by contrasting the brilliant experimental discoveries of the English natural philosopher with the theoretical prowess of the French savant. A. R. Hall reflected this interpretative tradition in his textbook on the *Scientific Revolution*, when he wrote in 1954 that Lavoisier was "less the author of new experiments than the first to realize their full significance."[3] Meldrum believed that "Priestley could produce new observations and discoveries in profusion. Lavoisier could not emulate Priestley; his strength lay in another direction."[4] These sentiments have lingered on in the images of the two scientists portrayed in more recent writing. They are, however, deeply colored by the discovery Priestley made of a specific air that became critical to Lavoisier, and by the historical success of Lavoisier's theory of combustion. If we look at the sustained laboratory practices of the two men, we can see that both were outstanding and innovative experimentalists, but in different ways. Neither could match the great chemists of their day, such as Andreas Marggraf, Wilhelm Scheele, or Torbern Bergman, in the employment of the classical qualitative methods of eighteenth-century chemistry to discover new acids, bases, or salts. Each developed a new genre of chemical experimentation. Priestley almost single-handedly invented the field of investigation of "airs" that prompted Lavoisier's comment that to Priestley one owed "facts which seemed to discover a new order of things." Lavoisier invented a quantitative style of chemical experimentation whose discoveries were of a different order. By the end of 1773 its shape was clear, but its potential was only beginning to unfold.

The contrast between the title of Lavoisier's Easter memoir proposing a "new theory" of calcination, and the title of part 2 of his *Opuscules* presenting "new research" on the existence of an elastic fluid symbolizes well the change that he had undergone in eight months. With the first title he had appeared as the discoverer of a theory so important that he could no longer defer its announcement, and as a prophet of a coming revolution. With the second he appeared as a patient experimentalist who had spent the last ten years making useful contributions to diverse areas within physics and chemistry. To those to whom he sent his book, he represented his own work on the elusive elastic fluid as a small addition to a theory originated by others. Were these merely changing poses that he found it convenient to adopt under changing circumstances? Lavoisier was so attentive to his image that some element of deliberate manipulation can be seen in much of his rhetoric. The change in his outlook about his position was, however, a realistic response to his experiences during the intervening months.

The older, heroic image of great scientists required them to be seen excelling their contemporaries from the outset. That expectation is well exemplified in the way Marcellin Berthelot characterized in 1890 the experimental venture that Lavoisier began in February 1773:

> In the solitary meditations of his laboratory Lavoisier formed the project of an enterprise whose character and scope he perceived from the beginning—we can cite in this regard the pages written and dated in his hand in 1773 in his experimental registers—and he achieved his enterprise with a method, a continuity, an invincible logic, while employing in due measure in the pursuit of his plan the facts already known and the particular discoveries that a group of men of genius, his contemporaries, were making every day, equally skillful experimentalists, perhaps more original than he in detail, but whose minds were less powerful.[5]

One century later we are less inclined to believe that men and women of extraordinary genius appear on the scene already fully endowed with the wisdom of their ultimate achievements. Studies of creativity show that in all areas, of the arts as well as the sciences, even the most gifted require years of practice to produce work of genuine originality. Lavoisier was no different. Like other mortals, he had to feel his way along, one step at a time, learning from his mistakes, benefiting from the successes of others, emerging only gradually as the leader in a busy field of activity that he set out to make his own. The first six months of his grand program of research on the processes that fix or release airs were more troubled than he ever revealed, or that historians have noticed. Of the many experiments he tried, nearly as many failed as worked. He did not come to these problems with a ready-made method or a clear view of the bumpy road ahead of him. The experience he gained over those months left him with a repertoire of apparatus and procedures far more effective for his purposes than the ordinary chemical equipment and operations with which he had begun. More important, he acquired a feel for what his techniques could do, and for their limitations. Without this repertoire of apparatus, procedures, and experience, what has been called his balance sheet method was no more than an idea. With them, the method was capable of bringing to material reality the changes which had been before then only the play of his imagination. But the pathway stretching out between him and the realization of his overarching dreams still remained long and twisting, probably far more so than he could foresee.

In a study of *Creating Minds* based on the biographies of seven towering figures of "the modern era," Howard Gardner has found, unexpectedly, that each of them required, at the critical period of their creative efforts, "both affective support from someone with whom he or she felt comfortable and cognitive support from someone who could understand the nature of the breakthrough."[6] Lavoisier did not carry through the formative phase of his venture in the "solitude of his laboratory." He, too, received emotional and intellectual support at this critical juncture. The dedication of his *Opuscules* to Trudaine de Montigny may have been politically advantageous to Lavoisier, but, as the foregoing account shows, it was also a genuine acknowledgment of what he owed to an older friend who had the insight to recognize the importance of what he was attempting, and the generosity to help him realize his aims.

In his letter to Lavoisier in August 1773, Trudaine contrasted his own status as one of the "amateurs" who only "love science," with those "who cultivate the sciences and who advance them."[7] As an Academician, Lavoisier was *expected* to advance science. As an Academician elected at a young age with no major scientific advances to his credit, he was probably more eager than many of his colleagues to prove that he belonged to this class of scientists. Some of his behavior during the spring of 1773 can be attributed to the eagerness of this ambitious young Academician to live up to such expectations. After occupying himself for several years with projects at which he did well, but that led to nothing extraordinary, he made, in the fall of 1772, a discovery that he saw at once as his best opportunity to become one of those who truly did advance the sciences. That he seized on that opportunity before he knew how hard it would become to exploit it fully, can thus be explained in part by the culture of the Academy to which he belonged. But his experience is more universal. It is shared by men and women of talent, in many times and places, who aspire to creative achievement, and who set for themselves challenging goals long before they can see how they will reach their destinations.

* CHAPTER TWELVE *

Before and After 1773

LAVOISIER STUDIES

THE preceding chapters have provided a concentrated narrative about one crucial year in the scientific life of a man whose total life was richly multifaceted. I have tried to present this episode as a story that is meaningful within its own boundaries, but that also offers new perspectives on what came before and after. It is time now to reflect tentatively on how this story can be connected with the broader pictures of Lavoisier that have emerged from a long tradition of historical scholarship, and with some of the recent trends in the historiography of the history of science.

I have drawn on laboratory notebooks to probe into the fine structure of the private science of a young Antoine Lavoisier in ways that I hope also illuminate the public figure that he was already becoming. Not all historians of science value such an approach. In a review of Gerald Geison's book *The Private Science of Louis Pasteur*, Robert Kohler has called laboratory notebooks a "trap" that "can confine our vision . . . to the man when we need to see him as exemplifying a distinctive scientific community and culture." According to Kohler, "The notebooks direct us to the private science of Pasteur when the really big story is his public science."[1] Does Kohler speak here for the dominant priorities of the discipline? In an age in which "shaping one's image" is often given priority over inner integrity, in which we talk about the "posture" someone "assumes" where formerly we discussed the "position one takes"; when we write about the "actors" or "players" in historical events rather than the "participants," it might seem that the public appearance is all that counts, that we have no interest in the inner person behind the personality we see on the stage or the screen. But I am not ready to conform to Kohler's hegemonic prescription for what historians of science should do.

The public scientist is important, and we must elucidate the ways in which scientists have assumed their public roles; but privacy is still valued, and the private science of those whose public performances have made them memorable is still worth pursuing. I believe that we still care about the person behind the mask, that we still seek the reality behind the appearance—not the transcendent reality of eternal truths, but the reality accessible to ordinary people who try to understand those around them and those who lived before them as deeply individual human beings. Lavoisier and Pasteur were both skilled public performers, but they were not hollow figures. Each had an inner character, of which he has left sufficient traces to permit meaningful interpretation.

Public and private are relative terms, not absolute dichotomies. In the semiprivacy of the laboratory in which he worked, before taking up quarters in the Arsenal that moved even his personal experimental trajectory into a public institution, Lavoisier seldom worked in isolation. His interactions with Trudaine de Montigny and with the other commissioners charged to referee his experiments typify a gregarious tendency to work in association with other scientists that marked Lavoisier out from his earliest years. Later on he initiated collaborative investigations that Roger Hahn has characterized as the germinal examples of the modern research team.[2] Moreover, his scientific style was shaped in part by the traditions and practices of the Academy of Sciences which he entered while still in the formative stages of his career. Bensaude-Vincent has written that

> an ensemble of implicit and explicit rules, embedded in an institution, the Royal Academy of Sciences of Paris, characterized a collectivity in a particular epoch, and organized the comportment of a member such as Lavoisier. . . . The impression of a research program pursued methodically should also be reconsidered within the framework of the habits and customs of the Academy.[3]

Bensaude-Vincent has described some of these rules and customs. The research program that Lavoisier began in 1773, and that I have treated here mainly as a personal venture, should also be reexamined as an exemplar of the impact of those institutional norms on the comportment of one of its rising stars.

By constructing the narrative of Lavoisier's emerging research program as a sequel to the "crucial year" previously explored by Henry Guerlac and others, I have somewhat insulated the story from the development of Lavoisier's scientific interests before 1772. There is reason to regard his move into the study of the "fixation and release of airs" as a fresh start. It came after several years in which, as his early biographer, Édouard Grimaux, noted, his travels, first on geographical surveys with Jean Etienne Guettard, and then as a member of the Tax Farm, had kept him away from Paris for long stretches of time. Only after his return from an inspection tour in February 1771, "could he take up again his work in the laboratory." He then busied himself with a variety of problems that were continuations of his prior interests, before finding the problem that began to dominate his scientific life eighteen months later.[4]

Despite this break in Lavoisier's scientific activities, historians have uncovered in his publications and manuscripts written during the 1760s, roots of some of the ideas and methods that became embedded into the research program begun in 1773. The most important such find has been J. B. Gough's identification of the "origins of Lavoisier's theory of the gaseous state" in a brief essay that Lavoisier wrote in 1766 as a commentary on an essay on the elements by Johann Theodor Eller.[5] Most recently, Marco Beretta has drawn attention to two other early documents—an outline for a chemistry course purported to have been written by Lavoisier and an unknown second person in 1764, and a manuscript of a treatise by Georg Ernst Stahl heavily annotated by Lavoisier and perhaps purchased by him in 1766—that may reveal further sources for Lavoisier's interest in the problems he began to investigate systematically only late in 1772.[6]

Some recent interpreters of Lavoisier's role in the chemical revolution have sought to identify in these early years the sources, not only of specific ideas that became important in his later research program, but also of broader "themes that infused his scientific achievements with coherence and meaning."[7] As I have already noted in chapter 1, several scholars, including Arthur Donovan, Evan Melhado,[8] and Marco Beretta,[9] have fixed on Lavoisier's admiration for physics as a *leitmotif* for his scientific endeavor, and have viewed his work in chemistry either as that of a physicist using chemistry as a site for the application of physical concepts, or as a chemist seeking to make his science more like physics. The roots of this presumed lifetime orientation can be located in his earliest scientific education, when Lavoisier studied methematics and attended lectures on experimental physics by the Abbé Nollet. Bensaude-Vincent has given a probing critique of these interpretations, drawing attention to the broad range of meanings of "physics" during the period of Lavoisier's scientific apprenticeship.[10] A fundamental criticism can also be made of the general historiographical procedure that extracts from the panoply of Lavoisier's earliest scientific views and experiences, and magnifies just those facets which appear to connect most strongly to his later endeavors. The result may be to create a semblance of greater continuity, singleness of purpose, and foreknowledge of the road ahead than Lavoisier, in his early scientific career, actually possessed.

Some biographies of Lavoisier quote a comment about the young Lavoisier by his mentor in geology, Guettard, that is very revealing: "I was accompanied on this journey [in 1763] by Lavoisier the son, whom a natural appetite for the sciences is leading to want to learn about all of them, before fixing himself on one or the other."[11] Through the next several years, Lavoisier continued to explore widely in the sciences of his time: in geology, mineralogy, meteorology, botany, and chemistry, as well as geometry and physics. In 1767 he viewed himself as engaged in the full range of the activities of natural history. It is audacious to mine the documents of this period only to explain his activities during the next decades, or to assume that during this period he had "fixed" on any one of these areas as the arena in which he would achieve future eminence. The documentation for Lavoisier's scientific activities during the 1760s is as rich as for any later period of his career. Not until historians have reconstructed these activities independently of the portents of later views and investigations that can be sifted from them,[12] will we be able to understand more deeply the connections between his scientific goals, ideas, and methods, before and after he started out on the momentous investigative pathway whose first phase is portrayed in the present volume.

The implications of this detailed account of Lavoisier's experimental activity in 1773 for our understanding of his later career can also be sketched out only in a preliminary way, until similar accounts of the succeeding episodes in the investigative journey on which he embarked in 1773 have been constructed. In my previous book on Lavoisier I have reconstructed some of these later episodes, but my focus on the "chemistry of life" caused me to describe in least detail those segments of his work most central to the theoretical claims around which the chemical revolution revolved. Most urgent, I believe, is to continue the full reconstruction through the years 1774–78, which culminated in his general theory of combustion and his first strong challenge to the phlogiston theory.

Lavoisier's characteristic balance sheet method emerged, according to my interpretation, in his experimental practice during the year

1773, as he first emulated, then superseded the methods of his two principal experimental models, the work of Stephen Hales and Joseph Black. Nowhere during this time or later, so far as I have found, did Lavoisier express verbally that he had during this period come to understand the foundations for his emerging method. Charles Gillispie has written that Lavoisier did not formulate the law of conservation of mass. "He assumed it. It was for him . . . a precondition but no finding of his science."[13] I agree with Gillispie, but with the qualification that Lavoisier had to learn to practice a science based on such an implicit "law," before the latter became self-evident to him.

Such a conclusion may fit recent trends in the history of science that have emphasized experimental practice and diminished the priority formerly given to theory, or to "scientific thought." The application of this shift to discussions of the chemical revolution is best represented in Jan Golinski's treatment of the debates between Lavoisier and Priestley and their respective partisans in his perceptive study of *Science as Public Culture.* "The participants came to appreciate," Golinski shows, "how debates about the facts are essentially debates as to how scientific practice is to be carried on." The resistance of British chemists to Lavoisier's chemistry was not, in Golinski's view, merely that of defenders of a theory under attack, but was an indication of "how different his experimental . . . practices were from traditional expectations."[14]

Like many others who have recently discussed experimental practice, Golinski is more interested in the public than the private face of that activity. He views experiments and the instruments with which they are performed as "instruments of persuasion," and explores their roles in "demonstration, authority, and community." Noting my use of manuscripts to examine the assumptions that Lavoisier and Laplace made in the performance and interpretation of the results of their experiments with the calorimeter, Golinski comments that his "purpose here is rather different: I am not concerned to judge the propriety of Lavoisier's interpretive maneuvers, but to record the degree of their persuasive success and failure and the objections made to them at the time."[15]

The question to be debated among historians of science is whether these two purposes are complementary, or whether Kohler is right to insist that the "big story" is that of the public scientist, and that those of us who attempt to illuminate the private science through laboratory records and other intimate documents are afflicted with narrow vision.

The most widely cited exemplar of the "new" histories of experimental practice, Steven Shapin and Simon Schaffer's *Leviathan and the Air Pump*, grounds its discussion of Robert Boyle's practice on a description of two experiments with the air pump isolated from the other forty-three experiments described by Boyle in his *New Experiments Touching the Spring of the Air*.[16] That is a slender foundation on which to raise the superstructure that they erect to depict the strategies Boyle used to "compel assent" to the "matters of fact" that his experiments yielded.[17]

Hans-Jörg Rheinberger has recently argued persuasively that the individual unit of scientific practice is not the single experiment, but the "experimental system." Once established, experimental systems have a "life of their own," which lead the investigators who wield them often toward novelties that they cannot anticipate in advance.[18] Whether we give priority to the system as Rheinberger does, or to the intentions of the investigator as I have tended to do in my explorations of "investigative pathways," we can draw the same conclusion: that it is necessary to reconstruct extended series of experiments to understand the dynamic of the creative practice of experimental science.

Lavoisier's experimental trajectory, extending over three decades before it was abruptly cut off, provides one of the best documented and most instructive examples of a successful experimental system, or investigative pathway, available to historians of science. I have explored some facets of this trajectory, but the subject is too rich to be exhausted by a single historian. I am hopeful that others will join in the further exploration of the treasures still hidden in the thirteen volumes of laboratory notebooks kept between 1773 and 1789, and the innumerable associated documents that Lavoisier left for us. I am also hopeful that those who portray the public face of Lavoisier's scientific persona and those who search for the inner dimensions of his person and his science, will not follow separate tracks, but will interact intensely, in mutual appreciation of the interdependence of these respective sides of one magisterial scientist.

I would particularly like to invite scholars of the chemical revolution to ponder the significance of the reinterpretation offered in chapter 10 (pp. 137–38) of Lavoisier's famous private remark, in February 1773, that the ideas he had just summarized in his laboratory register would occasion a revolution in physics and chemistry. Generations of Lavoisier scholars have treated this passage as a remarkably prophetic statement of the revolution he would later achieve. Bensaude-Vincent has shown, however, that the meaning of the chemical revolution, for Lavoisier himself, as well as for his followers and his opponents, continually changed as events unfolded.[19] Scholars have long debated whether Lavoisier originated a revolution or completed a revolution begun by his predecessors. That Lavoisier may have conceived *himself* in 1773 only as adding an increment to a revolution prepared by Joseph Black and others should, at the least, induce us to reexamine the meaning of his later statements as well. As his pathway after that year led him to the theory of combustion that had eluded him in 1773, as he challenged the Stahlians and confronted strong resistance to his views, he naturally came to attribute to himself a larger role in the revolutionary events of which he had assumed the uncontested leadership. But I believe that he never lost the sense that he was continuing what others had begun. In a letter to Joseph Black in 1790, he wrote, "It is very appropriate, Monsieur, that you should be informed of the progress which is being made in a pathway which you have opened up, and in which we all regard ourselves as your disciples."[20] According to the widely accepted view that Lavoisier had aimed from the beginning to launch a chemical revolution on the basis of his own initial discoveries about combustion and calcination, this passage appeared to be manufactured diplomacy, perhaps a gesture of reward for the recent conversion of Black to the new "French chemistry." It now appears, however, completely consistent with the interpretation of the famous passage in Lavoisier's laboratory notebook that I have suggested. This example well illustrates, I believe, that careful attention to the details of the private pursuit of science can reshape our understanding of the public meaning of that science.

* *Appendix* *

NAMES, SUBSTANCES, AND APPARATUS IN EIGHTEENTH-CENTURY CHEMISTRY

THE three common mineral acids of eighteenth-century chemistry, vitriolic acid, nitrous acid, and marine acid, were named for their sources (and were for that reason also known respectively as spirit of vitriol, spirit of niter, and spirit of sea salt). Concentrated nitrous acid was also known as aqua fortis, or, in French, *eau forte.* During the 1760s a fourth acid was distinguished from these, and named phosphorous acid. In the new nomenclature of Lavoisier and his associates, these four acids became nitric, sulfuric, muriatic, and phosphoric acid. After Lavoisier's oxygen theory of acidity had been replaced by the theory of hydracids in the early nineteenth century, muriatic acid was again renamed, becoming hydrochloric acid.

Traditionally, two forms of alkali were distinguished according to their behavior during a distillation. That which "flew up" and passed over into the receiver was called volatile alkali. That which remained in the distillation vessel was called fixed alkali. In 1736, Duhamel du Monceau demonstrated that the fixed alkali obtained from sea salt was distinct from that obtained from potash. The two were afterward designated, respectively, fixed mineral alkali and fixed vegetable alkali.

All three alkalis were known to exist in a mild form and a more caustic form. Joseph Black explained the difference between the two forms as due to the presence of fixed air in the mild form. In the new nomenclature, fixed air became carbonic acid, and the three mild alkalis were named, respectively, ammonium carbonate, potassium carbonate, and sodium carbonate.

The alkaline earths (or absorbent earths) were bases which, like the alkalis, formed neutral salts with acids. For Lavoisier's early investigations, the most important of these was known in its mild form as calcareous earth, which was usually either chalk or limestone. The caustic form was, when dry, called lime or quicklime; when saturated with water it was known as slaked lime. In the new nomenclature, calcareous earth became calcium carbonate.

Of the many neutral salts known to eighteenth-century chemists, only two figured repeatedly in the experiments of Lavoisier recounted here. Niter was composed of nitrous acid and fixed vegetable alkali. In the new nomenclature this salt became potassium nitrate. Sal ammoniac, a combination of volatile alkali and marine acid, was afterward known as ammonium chloride.

The product that resulted from heating a metal until it lost its metallic properties was called a calx of the metal. More generally, a calx was any powdery substance yielded by heating a substance strongly in air. Metals were usually obtained by heating their calces or ores mixed with charcoal, a process known as reduction. The ores from which metallic lead could be procured in this way were called litharge (yellow lead) and minium (red lead).

Early in the eighteenth century, Georg Ernst Stahl named the "inflammable principle," thought to be transferred from one substance to another in such processes, phlogiston. Present not only in metals, but in combustible bodies such as phosphorus, sulfur, fats, oils, wax, and wood, phlogiston was thought to be driven off from these substances in combustion or calcination. By suggesting that the same identical substance was transferred in all these operations, that reductions resulted from the restoration of phlogiston to the metals from the charcoal, and that the reduction of vitriolic and other acids to sulfur or an analogous substance also involved the addition of phlogiston, Stahl provided a qualitative unifying framework that encompassed many of the most important operations of contemporary chemical practice.

The new theory of combustion that Lavoisier proposed in 1778 replaced these conceptions. Where phlogiston had formerly been released, oxygen was now absorbed. The metallic calces were afterward called oxides.

The volatile liquid and solvent obtained by distilling wine and other spirits, customarily called spirit of wine in the eighteenth century, was renamed alcohol in the new nomenclature. A solvent was sometimes called a menstruum (plural menstrua), and an insipid, watery fluid was sometimes called phlegm.

Fuller descriptions of eighteenth-century chemical nomenclature are given in Jon Eklund, *The Incompleat Chymist* (Washington, D.C.: Smithsonian Institution Press, 1975), and in Maurice P. Crosland, *Historical Studies in the Language of Chemistry* (London: Heineman, 1962).

In many of the experiments Lavoisier performed in 1773, he used or adapted apparatus traditionally used for distillations. The vessels in which substances were heated in the furnace were given different names, depending on their shapes. A spherical vessel with a neck curving to the side was called, then, as now, a retort. A spherical vessel with a long, narrow, vertical neck was known as a matras. A matras was often fitted with a separate "head" connected to the downward slanting "spout," "nose," or "beak." A distillation vessel whose shape resembled that of a gourd was known as a cucurbite. Receivers of various sizes and shapes were used. When large and spherical they were called, in French, "ballons."

Lavoisier utilized the most common French premetric system of weights, in which:

1 pound = 16 ounces
1 ounce = 8 gros
1 gros = 72 grains

The measurements of length Lavoisier used in the experiments described were:

1 inch (*pouce*) = 1.06 English inches
= 12 lines (*lignes*)

* Notes *

NOTE to the Reader: The passages quoted in French from Lavoisier's laboratory registers retain in most cases the irregular spellings, capitalizations, and punctuations found in the original.

INTRODUCTION

1. Lavoisier's career can be viewed as a prime example of what Robert Merton has called the "process of cumulative advantage" enjoyed by persons who are recognized early as exceptionally talented by older mentors or sponsors influential enough to elevate them quickly into elite positions which provide them extraordinary opportunities to develop their talents, and high visibility for their subsequent achievements. Robert K. Merton, "The Sociology of Science: An Episodic Memoir," in *The Sociology of Science in Europe*, ed. Robert K. Merton and Jerry Gaston (Carbondale: Southern Illinois University Press, 1977), pp. 89ff.

2. See, for example, Henry Guerlac, *Antoine-Laurent Lavoisier: Chemist and Revolutionary* (New York: Charles Scribner's Sons, 1975) (published also as the entry "Lavoisier, Antoine-Laurent," in *Dictionary of Scientific Biography*, ed. Charles Gillispie (New York: Charles Scribner's Sons, 1970–80), pp. 66–91; Arthur Donovan, *Antoine Lavoisier: Science, Administration, and Revolution* (Oxford: Blackwell, 1993); Jean-Pierre Poirier, *Antoine Laurent de Lavoisier 1743–1794* (Paris: Pygmalion, 1993). The first biography of Lavoisier based on his unpublished documents, Édouard Grimaux, *Lavoisier 1743–1794* (Paris: Félix Alcan, 1888), has recently been reprinted.

CHAPTER ONE
THE SOURCES OF LAVOISIER'S QUANTITATIVE METHOD IN CHEMISTRY

1. Antoine Lavoisier, *Traité élémentaire de chimie* (Paris: Cuchet, 1789), 1:140–41.

2. Richard E. Dickerson, Harry B. Gray, Marcetta Y. Darensbourg, and Donald J. Darensbourg, *Chemical Principles*, 4th ed.(Menlo Park, CA: Benjamin Cummings, 1984), p. 36.

3. Lavoisier, *Traité*, 1:141.

4. Marcellin Berthelot, *La révolution chimique: Lavoisier* (Paris: Alcan, 1890), p. 41.

5. Emile Meyerson, *Identity and Reality*, trans. Kate Loewenberg (New York: Dover Publications, 1962), p. 157.

6. Henry Guerlac, *Lavoisier—The Crucial Year: The Background and Origin of his First Experiments on Combustion in 1772* (Ithaca: Cornell University Press, 1961), p. xviii.

7. Lavoisier, *Traité*, 1:141.

8. John G. McEvoy, "Continuity and Discontinuity in the Chemical Revolution," *Osiris* 4 (1988): 203–5.

9. Bernadette Bensaude-Vincent, *Lavoisier: Mémoires d'une révolution* (Paris: Flammarion, 1993), p. 210.

10. Ibid., p. 213.

11. Charles Coulston Gillispie, *The Edge of Objectivity* (Princeton: Princeton University Press, 1960), pp. 231–32.

12. Guerlac, *Crucial Year*, pp. 11–24.

13. See, for example, Max Speter, *Lavoisier und seine Vorläufer* (Stuttgart: Ferdinand Enke, 1910), p. 45.

14. Donovan, *Antoine Lavoisier*, pp. 45–73.

15. J. Heilbron, "Introductory Essay," in *The Quantifying Spirit of the Eighteenth Century*, ed. Tore Frängsmyr, J. L. Heilbron, and Robin Rider (Berkeley: University of California Press, 1990), pp. 1–23.

16. Donovan, *Antoine Lavoisier*, p. 49.

17. Guerlac, *Crucial Year*, p. xix.

18. C. E. Perrin, "Document, Text, and Myth: Lavoisier's Crucial Year Revisited," *British Journal for the History of Science* 22 (1989): 3–25.

19. "Cahier de laboratoire de Lavoisier," Numero 1 (henceforth, "Cahier lab. 1") Archives de l'Académie des sciences, Paris, p. 1; reproduced in Guerlac, *Crucial Year*, pp. 228–29.

Chapter Two
Consequences of the Crucial Year

1. Lavoisier, "Mémoire sur lacide du phosphor et sur ses combinaisons avec differentes substances salines terreuses et metalliques," Lavoisier papers, fol. 1308D, AAdS, reproduced in Guerlac, *Crucial Year*, pp. 224–27.

2. Perrin, "Document," p. 16.

3. Lavoisier, "Sur la cause de laugmentation de pesanteur quaquierent les metaux et quelques autres substances par la calcination," transcribed in C. E. Perrin, "Lavoisier's Thoughts on Calcination and Combustion, 1772–1773," *Isis* 77 (1986): 664.

4. Guerlac, *Crucial Year*, pp. 125–38.

5. Lavoisier, "Cause de laugmentation," in Perrin, "Lavoisier's Thoughts," p. 664.

6. Lavoisier, "The Sealed Note of November 1, 1772," in Guerlac, *Crucial Year*, pp. 227–28.

7. Perrin, "Lavoisier's Thoughts," p. 658.

8. Stephen Hales, *Vegetable Staticks* (New York: Science History Publications, 1969), p. 163.

9. Lavoisier, "Cause de laugmentation," in Perrin, "Lavoisier's Thoughts," p. 664.

10. "Cahier lab. 1," pp. 1–4.

Il est certain quil se degage des corps dans un grand nombre de circonstances un fluide elastique mais il existe sistemes sur sa nature. les uns comme Hales et ses sectateurs ont pensé que c'étoit l'air lui meme celui de l'atmosphere qui se combinent avec les corps soit par des operations de la vegetation et de l'economie animale soit par des operations de l'art. il n'a pas presumé que le fluide peut etre different de celui que nous respirons a la difference quil est plus chargé de vapeurs nuisibles ou bienfaisants suivant la nature des corps dont il est tiré. quelques uns des Phisiciens qui ont suivis M. Hales ont remarqué des differences si grandes entre lair degagés des corps et celui qui nous respirons quils ont pensé que cetoit une autre substance a la quelle ils ont donné le nom dair fixe. un troisième ordre de Phisiciens ont pensé que la matiere elastique qui s'echappe des corps etoit different suivant les substances dont il avoit ete tiré et ils ont conclu que ce netoit qu'une emanation des parties les plus subtiles des corps dont on pouvoit distinguer une infinité d'especes. . . .
il est constant que lair fixe present des Phenomenes tres different de l'air ordinaire en effet il tue les animaux qui le respire tandis que celui cy est essentiellement necessaire a leur conservation. il se combine avec un tres grande facilité davec tous les corps tandis que lair de l'atmosphere dans les memes circonstances se combine avec difficulté et peut etre meme ne se combine du tout. . . .
. . . de tout repeter avec de nouvelles precautions affin de lier ce que nous connoissons sur lair qui se fixe ou qui se degage des corps avec les autres connoissances des corps et . . . et former une theorie. . . . il reste une suitte dexperiences immense a faire. . . .

un point important que la plus part des auteurs ont negligé cest de faire attention a l'origine de cet air qui se trouve dans un grande nombre de corps. ils auroient appris de M. Hales quune des principles operations de leconomie animale et vegetale consiste a fixer lair a le combiner avec leau et le feu et a forme la terre et a former tous les combinés que nous connoissons. ils auroient encore vus que l'air le fluide elastique qui sort de la combination des acides soit avec les alkalis soit avec toute autre substance vient encore originairement de l'atmosphere dou ils auroient ete en etat de conclure ou que cette substance est lair lui meme combiné avec quelque substance partie volatile qui s'emane dun corps ou au moins que cest une substance extraitte de lair de l'atmosphere.

Cette facon denvisager mon objet m'a fait sentir la necessité de repeter dabord et de multiplier les experiences qui absorbent de l'air affin que connoissant lorigin de cette substance je puis suivre ses effets dans toutes les differentes combinaisons.

Les operations par lesquelles on peut parvenir a fixer de l'air sont la vegetation la respiration des animaux la combustion dans quelques circonstances la calcination, enfin quelques combinaisons chimiques. Cest par ces experiences que j'ay cru devoir commencer.

11. Lavoisier, "Sur la cause de laugmentation de pesanteur quacquierent les metaux et [plusieurs] quelques autres substances par la calcination," Fonds Lavoisier, fol. 1303, AAdS; reproduced in Perrin, "Lavoisier's Thoughts," pp. 662–65.

12. See Perrin, "Lavoisier's Thoughts," p. 654.

Chapter Three
Vision and Reality

1. Joseph Black, *Experiments on Magnesia Alba, Quicklime, and some other Alkaline Substances* (Edinburgh: Alembic Club, 1944), p. 25.

2. Lavoisier, *Opuscules physiques et chymiques* (Paris: Durand, 1774), pp. 217–18.

3. Cahier lab. 1, pp. 7–9. "j'ay vu avec etonnement quelle faisoit une vive effervescence." "J'ay voulu faire chauffer pour eprouver si lair se combineroit plus aisement, mais le matras a cassé."

4. Black, *Magnesia Alba*, p. 22.

5. Lavoisier, "Cause de laugmentation," in Perrin, "Lavoisier's Thoughts," p. 664.

6. Cahier lab. 1, pp. 10–11.

J'ay pris 2 livres de plomb bien pur je l'ay couté en lames fort minces que j'ay divisé avec des ciseaux. je les ay introduit dans une cornue de gres et j'y ay appliqué lappareil de M. Hales. j'ay ete lentement ensuitte j'ay poussé le feu vivement pendant quatre heures. les barres qui soutenoient la cornue etoient absolument rouges. l'air s'est dilaté pendant toute lopperation mais ce qui m'a le plus inquieté cest que meme lorsque le feu a en acquis son grand degre leau baissoit toujours. j'ay arreté et bientot lair a commencé a se condenser et l'eau est remonté. j'ay ete tres etonné quand au bout de 2 heures je l'ay vu [. . .] et bientot redescendre au lieu de monter. verriffication faitte de la cornue elle avoit une petite fente imperceptible par ou l'air s'etoit introduit.

. . . je suis encore tombé a une cornue qui prenoit air.

7. Ibid., pp. 11–12.

Ces accidens repetés m'ont obligé de recourir aux cornues de verre. . . . je l'ai augmenté malgre moi et j'ay vu l'eau recommencer a descendre ce qui ma fait craindre que la cornue ne fut rentrée sur elle meme et cest qui etoit arrivé en effet elle etoit entierement deformée en refroidissant meme elle s'est fetée et il est entré de lair ainsy experience manquée.

8. Maurice Daumas, *Lavoisier: théoricien et expérimentateur* (Paris: Presses Universitaires, 1955), p. 29.

9. Cahier lab. 1, pp. 12, 12v.

10. Ibid., p. 12v.

cette derniere Substance est suivant moi une Combinaison dair et de plomb. Cet air devoit passer dans lalkali volatil avec lequel il a une prodigieux affinité. cependent il se dissipe et s'echappe pendant l'operation.

Cette difficulté est embarassante et je crois quon doit conclure que l'air combiné avec le plomb dans le minium nest point cet air fixe qui a tant daptitude a se combiner avec les corps alkalis cest sans doute lair meme de latmosphere. peut etre aussy cet air qui se degage du plomb nest-il pas assez chargé de phlogistique pour se combiner avec les alkalis. Car suivant le sisteme de quelques uns lair fixe est un air combiné avec le phlogistique mais javoue que tout cecy presente beaucoup dincertitude.

11. Cahier lab. 1, p. 13.

12. Ibid., p. 14.

un premier point dans les experiences sur lair fixe est de determiner son poids. il sagira ensuitte d'examiner s'il est elastique comme celui de latmosphere en troisieme lieu jusqu'a quel point il est compressible. . . . 4. de le laver avec differentes matieres pour observer ensuitte son influence sur les animaux.

13. Ibid., pp. 14–15. "la difference des poids donnera le rapport de pesanteur de l'air fixe et de l'air ordinaire. . . . la difficulté sera de bien secher le ballon."

14. Ibid., pp. 16–17.

15. Ibid., p. 17. "dans les experiences sur les animaux ne pas obmettre les grenouilles."

16. Ibid., p. 18.

Les differentes machines cy dessus indiqués ne sont point encore finies par le lenteur des ouvriers une maladie d'ailleurs de quinze jours et differentes affaires m'ont obligé d'interrompre mes experiences Cependant comme je voudrais pouvoir annoncer quelque chose a la rentrée et que le tems presse jimagine quon pourroit faire une partie des memes experiences au verre ardent avec assez de simplicité.

17. Guerlac, *Crucial Year*, pp. 156–59; Lavoisier, *Opuscules*, p. 257.
18. Cahier lab. 1, p. 20.
19. Lavoisier, *Opuscules*, p. 256.
20. Ibid., pp. 18–19; Lavoisier, *Opuscules*, p. 192.
21. Ibid., pp. 19–20.

J'ay commencé des lors a soupconner que le contact dun air circulant etoit necessaire a la formation de ce chaux metallique. que peut etre meme la totalité de lair que nous respirons nentroit pas dans les metaux que l'on calcine mais seulement une portion qui ne se trouve pas bien abondamment dans une masse donnée dair. peut etre aussy la couche de chaux qui couvre la surface du metal empechoit-elle le contact direct de l'air et arretoit la progres de la calcination.

22. Ibid., pp. 17, 19–21. ". . . pres de 7 pouces dair. Cest a dire pres de quatre grain. . . ." "neanmoins cest une operation a recommencer." "il avoit au contraire perdu environs demi graine Ce qui vient sans doute des vapeurs qui s'en etoit exalées."
23. Ibid., p. 21.
24. Ibid., pp. 21–22. "je crains que le charbon n'y a contribué."
25. Ibid., p. 22. "mais cette experience ne prouve rien de bien direct par la raison que lalkali . . . qui etoit tres privé d'air et par consequent tres avide a pu absorber la portion qui setoit formée."
26. Ibid., p. 23.
27. Ibid., pp. 23–24; Lavoisier, *Opuscules*, pp. 285–87. "il s'est elevée beaucoup de fumée blanche dont partie s'est attachée a la partie superieure de la cloche partie s'est deposée sur la surface de lhuille."
28. Cahier lab. 1, p. 24.
29. Ibid., p. 25.
30. Ibid., pp. 26–27.
31. See, especially, Lavoisier, *Opuscules*, p. 258.
32. Hales, *Vegetable Staticks*, p. 119.
33. Ibid., pp. 28–29.

Je me suis fait bien de fois une objection tres forte contre mon sistem de la reduction metallique et voicy en quoi elle consiste la chaux suivant moi est une terre calcaire privée d'air les chaux metalliques au contraire sont des metaux saturés d'air cependant les uns et les autres produisent un effet semblable sur les alkalis, ils les rendent caustiques.

34. Ibid., p. 29.

> Lobjection me paroit complettement détruitt pour ce qui regarde lakali fixe mais elle subsiste en entier pour lakali volatil en effet lalkali volatil quon obtient de la combinaison du sel ammoniac et du minium est deliquescent il ne precipite pas l'eau de chaux. il est infiniment penetrant il est donc evident quil est dans un etat de causticité. Comment expliquer ce phenomene? j'avoue que je ne sais rien encore.

35. Ibid., p. 51.

Chapter Four
The Public Arena

1. Daumas, *Lavoisier*, p. 28.
2. Perrin, "Lavoisier's Thoughts," p. 658.
3. Reproduced in Guerlac, *Crucial Year*, p. 228.
4. Lavoisier, "Cause de laugmentation," in Perrin, "Lavoisier's Thoughts," p. 665.
5. Perrin, "Lavoisier's Thoughts," p. 655.
6. Lavoisier, "Sur une nouvelle theorie de la calcination et de la reduction des substances metalliques et sur la cause de l'augmentation de poids quelles acquerent au feu soit par la calcination soit par quels autres procedes analogues," Fonds Lavoisier, 1303, Archives de l'Académie des sciences, Paris.

> Lespece de fermentation qui regne dans presque tout leurope scavant sur la nature et les proprietes de lair fixe le grand nombre de brochures qui paroissent sur cet objet le grand nombre de scavans qui s'en occupent ne me permettent pas de defferer plus longtemps de faire part a lacadémie de quelques experiences que j'avois deposé dans son sein et dou je ne comptois faire usage que lorsque j'en auroit formé un corps de theorie complet.

7. Ibid. "quelques experiences que j'avois deposé il y a pres d'un an dans son sein et que je m'etois proposé de n'en faire paroitre que lorsque j'aurois ete entierement satisfait de mon travail."
8. Ibid. "le merite distingue de quelq'uns qui s'en occupent", "enfin la crainte de vois enlever a la chimie francoise et a moi meme des decouvertes que je crois importantes."
9. Ibid.

> la circonstance actuelle ne me permit pas de donner icy le detail de mes experiences je me contenterai de presenter le resultat je revendrai ensuitte dans nos seances particuliers sur [chacun de ces objets] ‹chaque objet en particulier› je donnerai le detail des [experiences] ‹operations› la description

des differents appareils, auxquels j'ay ete obligé davoit recours ainsy que des machines que j'ay employées. ‹je multiploient les preuves autant qu'il me sera possible.›

10. Lavoisier, "Sur une nouvelle theorie de la calcination et de la reduction des substances metalliques sur la cause de laugmentation de poids quelles acquirent au feu et sur differens phenomenes qui appartiennent a l'air fixe," in René Fric, "Contribution à l'étude de l'évolution des idées de Lavoisier sur la nature de l'air et sur la calcination des métaux," *Arch Int d'Hist. Sci.* 12 (1959): 156.

11. Lavoisier, "Nouvelle theorie," Fonds Lavoisier, 1303; Lavoisier, "Nouvelle theorie," in Fric, "Contribution," p. 156.

Si au lieu de faire les experiences a l'air libre on les fait dans une portion d'air enfermée sous une cloche de verre renversé dans une . . . jatte et qu'on intercepte la communication avec l'air de l'atmosphere par le moyen de l'eau de l'huille ou du mercure a mesur que ces metaux se reduisent en chaux le volume d'air diminue et l'augmentation de poids du metal se trouve a peu pres egale a la quantité d'air absorbé. Si par le moyen d'une verre ardent ou par quelque autre procedé dont je donnerai les details on parvient a faire la reduction de ces metaux cet a dire a les faire passer de l'etat de chaux a celui de metal aussitot ils restituent tout l'air qu'il avoient absorbé et reperdent en meme temps toute l'augmentation de poids qu'ils avoient acquise.

12. Lavoisier, "Nouvelle theorie," Fonds Lavoisier, 1303.

j'ay pris [1 gros d'etain et un gros] ‹parties egales de plomb et› detain que j'ay melé ensemble je les ay tenus long tems exposés au foyer du verre ardent [dans un appareil convenable] ‹sous une cloche de verre et dans une quantité d'eau donné› Je suis parvenu a les calciner asses bien. non pas cependant aussy promptement qu'a l'air libre. l'operation finie la quantité d'air absorbé s'est trouvée égale a 125 fois le volume des matières employées.

La reduction du minium et de la litharge m'a donné l'inverse de cette experience. on scait en générale que toute reduction metallique est accompagné d'une mouvement deffervescence cette remarque se trouve dans tous les traitte de chimie et de metallurgie. je me suis assuré par nombre d'experiences que cette effervescence etoit du a l'air fixe que s'en degage. je decrirai les appareils par le moyen desquels je suis parvenu a le retenir et a mesurer la quantité. C'est surtout sur le minium et la litharge que j'ai multiplié ces experiences on les reduit aisemen par une simple addition d'un douxieme de leur poids de poudre de charbon cette experience m'a toujours donné de [2 au 4] ‹au moins 3› cent fois son volume d'air a quelques variete pres qui tiennent a une circonstance dont je rendrai compte dans la suitte.

Additions in margin:

[je suppose quelques exceptions a ces regles generales que j'ay quelque fois observé et dont je rendrai raison dans une autre circonstance.]
‹si j'ay rencontré quelquefois des exceptions a ces regles generales elles tenoient a des circonstances particulieres dont je rendront raison dans une autre tems.›

13. Lavoisier, "Nouvelle theorie," in Fric, "Contribution," p. 158.
14. Lavoisier, "Nouvelle theorie," Fonds Lavoisier, 1303.

il resulte evidemment de ces experiences 1^0 qu une chaux metallique n'est autre chose que le metal lui meme combiné avec de l'air fixe 2^0 que la reduction metallique ne consiste que degage l'air des chaux metalliques Soit celui presentant un corps avec lequel il a plus d'analogie ou autrement. 3^0 que c'est a l air fixe contenu dans l'atmosphere que le metaux doivent l'augmentation de poids.

15. Ibid.

Cette theorie est destructive de celle de Stahl adopté par presque touts les chemistes et cette circonstance soit bien propre a me tenir garde je n'ay pu cependant ne refusera l'evidence surtout lorsque des experiences decisives m'ont assuré qu'il etoit possible de reduire presque tous les metaux sans addition de phlogistique je reviendra sur cet objet dans un autre memoire.

Correction in left margin:

Cette doctrine est directement contraire a celle [etablie] ‹enseignée› par Stahl et adopté depuis generalement par tous les chimistes. Cette circonstance . . . . etoit bien capable ‹de me donner de la méfiance› de me mettre en garde contre mes ‹propres› experiences.

16. Ibid.; Lavoisier, "Nouvelle theorie," in Fric, "Contribution," pp. 158–62.
17. Lavoisier, "Nouvelle theorie," Fonds Lavoisier, 1303.
18. Ibid.
19. Lavoisier, "Nouvelle theorie," in Fric, "Contribution," p. 158.
20. Ibid., p. 162.
21. Perrin, "Lavoisier's Thoughts," p. 661.
22. Lavoisier, "Nouvelle theorie," in Fric, "Contribution," p. 162.
23. Lavoisier, "Mémoire sur la combustion en général," *Mémoires de l'Académie Royale des Sciences*, 1777 [pub. 1780], p. 593.
24. Lavoisier, "Nouvelle theorie," in Fric, "Contribution," p. 162.
25. Guerlac, *Crucial Year*, pp. 41–42.
26. Ibid., pp. 55–58.
27. *Mercure de France*, May, 1773, pp. 173–74.
28. Donovan, *Antoine Lavoisier*, p. 103.

Chapter Five
Reflections

1. J. B. Gough, "Lavoisier's Early Career in Science: An Examination of Some New Evidence," *British Journal for the History of Science* 4 (1968): 52–57.

2. Lavoisier, "Reflexions sur l'air," in Fric, "Contribution," p. 145.

3. Lavoisier, "Essay sur la nature de l'air," in Fric, "Contribution," pp. 147–48.

4. Ibid., pp. 149–50.

5. Lavoisier, *Opuscules*, pp. 1–2.

6. Daumas, *Lavoisier*, p. 30.

7. According to Meldrum, Lavoisier's account ended originally with Joseph Priestley, and the last four chapters on French chemists, including Bucquet were added later, after the commission appointed to review Lavoisier's text objected to his omission of French chemists. That is a plausible interpretation of the structure of the text. On the other hand, as will be clear below, Bucquet's presentation influenced Lavoisier's own investigation at one point in the spring of 1773, during the time he was composing his text. See Andrew Norman Meldrum, *The Eighteenth Century Revolution in Science—The First Phase* (Calcutta: Longmans Green, 1930), p. 18.

8. Guerlac, *Crucial Year*, pp. 25–35.

9. Lavoisier, *Opuscules*, pp. 11–26.

10. Guerlac, *Crucial Year*, p. 29.

11. According to Meldrum, Lavoisier himself "had a hand in these publications." Meldrum, *Eighteenth Century Revolution*, p. 17.

12. Lavoisier, *Opuscules*, pp. 37–43.

13. Ibid., p. 109.

14. Ibid., pp. 109–70.

15. Ibid., p. 87.

Chapter Six
In the Shadow of Black

1. Lavoisier, *Opuscules*, pp. 263–65.

2. Cahier lab. 1, p. 31. "un leger abaissement de l'eau dans cette operation mais moindre que je ne le pensois."

3. Ibid. "mais je n'ay pas absolument pu lenflammer parce que restant toujours au fond il ne peut eprouver le contact de l'air qui lui est necessaire pour bruler."

4. For a discussion of the inevitability of unforeseen pitfalls in the early phases of development of any new form of experimentation, see J. R. Ravetz, *Scientific Knowledge and Its Social Problems* (Harmondsworth: Penguin Books, 1973), pp. 94–101.

5. René Fric, ed., *Oeuvres de Lavoisier: Correspondance* 2 (Paris: Albin Michel, 1957), pp. 389–90.

6. Lavoisier, "Nouvelle theorie," Fric, "Contribution," p. 159.

7. Cahier lab. 1, p. 33.

8. Ibid., pp. 33–34; Lavoisier, *Opuscules*, pp. 171–72. "Maniere dobtenir de l'air fixe sans vessie et par un appareil qui ne laisse aucun doute sur sa purité."

9. Cahier lab. 1, pp. 35–36; Lavoisier, *Opuscules*, p. 45. "Experience sur leffet dun refroidissement tres grand sur lair fixe."

10. Cahier lab. 1, p. 37. "Appareil pour faire passer lair quon veut Soumettre a lexamen soit a travers leau de chaux soit a travers telle autre liqueur qu'on voudra."

11. Ibid.

12. Ibid., pp. 38–39.

> Lalkali volatil tiré du sel ammoniac par le minium est vraiment un alkali caustique je l'ay epuisé avec leau de chaux et il ne la precipite aucunement Ce fait est extremement singulier et me paroit tout a fait inexplicable dans le principe de la fixation de l'air dans les chaux metalliques.

13. Ibid., pp. 39–40.

> Croyant quil bouilleroit aisement j'ay eu sur la fin quelques bouffées de vapeur qui ont fair varier la colonne de mercure mais je n'ay pu massurer s'il y avoit eu haussement ou abaissement j'ay repeté lexperience avec lesprit de vin chaud jusqu'au degre de lebullition. il ne m'a pas paru bouillir sensiblement plus vite sous le recipient j'ay eu de meme une bouffée debullition apres quoi la liqueur est resté tranquille j'ay repeté plusieurs fois lexperience et il ma paru constant que lesprit de vin ne boul pas dautant plus vite comme l'eau quil est chargé dune moindre colonne de latmosphere.
>
> je crois entrevoir ausy quil boul aussy aisement a froid qua chaud sous la machine pneumatique. . . .
>
> il ne paroit pas quelles fassent remonter le mercure.

14. Ibid., p. 41. "aucun vestige ni d'eau mere ni de sel marin" "Ce qui m'a paru tres extraordinaire car la craye avant cette operation etoit fort humide. est-ce que la craye calciné a un feu doux reprendoit de l'air fixe."

15. Black, *Magnesia alba*, p. 22.

16. Lavoisier, "Nouvelle theorie," in Fric, "Contribution," pp. 161–62.

17. Cahier lab. 1, pp. 48–50. "avec un appareil propre a mesurer lair. . . il ne parut pas se degager de fluide elastique". . . . "au bout de deux jours il y avoit production dair considerable."

18. Black, *Magnesia alba*, p. 24.

19. Cahier lab. 1, p. 42. "je n'ay employé reelement que 28 onces 6 gros de chaux. . . . qui pesoit tout evalué 37 onces justes. . . . restoit au chaudron a la Spatul, etc. . . . le rapport de la chaux vive a la chaux eteint et dessechée est

donc comme 1000 a 1287 autrement dite la chaux vive absorbe un peu plus dun quart d'eau."

20. Black, *Magnesia alba*, p. 26.
21. Ibid., p. 28.
22. Cahier lab. 1, pp. 42–44.
23. Ibid., pp. 44–45.

> la petite incertitude qui laissoit cette experience m'a engagé a la recommencer. . . . j'ay saturé en allant assez vite. . . . j'ay lieu de croire que la saturation etoit exacte . . . il a fallu dans cette experience 13.5 gros de Chaux eteinte pour saturer 48 gros dacide nitreux cest un peu plus du quart et je crois ces proportions tres exactes en supposant lacide nitreux 100 il faut un peu plus de 28.

24. Ibid., p. 45.

> j'ay fait le meme operation sur la craye cy dessus tres exactement dessechée dans une cuilliere de fer. jay employé la meme dose dacide nitreux il y a eu une vive effervescence. elle est si forte et la matierre se boursouffle tellement sur la fin quil faut une grand vase il y a eu environ 1 gros de perte qui a coulé pendant loperation tout au plus je ne levalu meme que 56 grains.

25. Ibid., p. 46. "sensiblement passé le point de saturation de 1 ou 2 gros. . . . le rapport est donc comme 22 a 48 ou comme 458 est a 1000. la diminution de poids a ete donc dune once ou 8 gros sur 22. cest a dire comme 364 a 1000."

26. Ibid. "il suit de ce cy que 22 gros de craye ne contiennent que 14 gros de terre calcaire et d'eau le reste est la fluide elastique."

Chapter Seven
Caution and Consolidation

1. Ibid., p. 46.

> j'ay mis sous la machine pneumatique de la chaux vive et de l'eau et j'ay fait le vuide la chaux sest eteinte comme a lordinaire sest reduitte en pulpe avec chaleur je ne me suis pas apperсu quil y eut de production ni dabsorbtion dair dans cette operation mais comme le recipient lutte a la machine pneumatique laissoit introduire de lair cette experience est a recommencer.

2. Ibid., p. 40.

> il ma paru que le mercure descendoit de quelque chose pendant lebullition mais de peu et surement il ne monte pas. Ce qui est fort singulier. il y a une evaporation considerable pendant cette operation et lesprit de vin se rassemble en gouttes sur le bocal ou est lesprit de vin lequel epreuve un refroidissement assez grand.

3. Ibid., p. 47.

> j'avois bien de crois dapres la theorie que cetoit une terre calcaire et non une chaux. pour m'en assurer je le laisse dans un grand nombre deaux mais je m'appercus quelle se dissolvoit et que cetoit da la chaux ce fait est singulier car lalkali caustique obtenu par cette operation fait lui meme un leger mouvement deffervescence avec les acides. ce cy demande a etre eclairci.

4. Ibid. "un tres beau precipitate bleu mais la terre calcaire n'a point ete precipitée."

5. Black, *Magnesia alba*, p. 27.

6. Cahier lab. 1, pp. 47–48. "enfin de lalkali fixe ordinaire a donné un precipite blanc un peu jaunatre a peu pres comme la soude."

7. Ibid., pp. 48–50. "rien nanoncoit lacidite et elle ne rougissoit point du tout le papier bleu."

8. Ibid., p. 51.

> mais lentement et beaucoup moins que par lair fixe qui sort de la craye en effervescence. Cependant je suis parvenu a avoir un leger precipite et il est probable quen continuant plus long tems j'aurois precipité toute la chaux.

9. Ibid., p. 52. "elle m'a paru s'en impregner peu." "Ce qui prouve que lair de latmosphere se mele en peu de tems avec lair infecte."

10. Lavoisier, *Opuscules*, pp. 109–50.

11. Ibid., pp. 259–60.

12. Cahier lab. 1, p. 53; Lavoisier, *Opuscules*, pp. 260–65.

13. Cahier lab. 1, pp. 53–54.

> Comme le lut commencoit beaucoup a s'echauffer il mest venu linquietude que la quantite dair qui etoit produitte ne vint de lexterieur j'ay resseré le lut en le tiant avec du fil en meme tems le feu baissoit et la production dair a cessé et le vaisseaux ont bien conservé le vuide le reste de la journée. mais j'ay toujours lincertitude de scavoir s'il nest point entré dair exterieur.
>
> la quantite de plomp [*sic*] qui devoit resulter de cette experience etoit environ de 3/4 de pouces cubes—dou il suivoit que le plomb en se convertissant en minium se charge de 388 fois son volume dair.

14. Ibid., pp. 54–55.

15. Ibid., p. 55. "il y avoit donc un perte de 38 pouces cubiques qui s'etoit combiné avec la chaux Ce qui repondoit a un poids dair de 18 a 20 grains en supposant lair fixe de meme pesanteur que celui de latmosphere."

16. Ibid., pp. 53v–54. "Table des volumes dair repondant a un certain poids de l'eau."

17. Ibid., p. 56. "j'ay voulu mettre plus dexactitude dans les resultats de mon operation du 23 may" "cest la quantite dair produitte" "ce qui est trop parce que

lon partiroit dapres un air dilaté" "la quantite dair degagé est denviron 2 gros quelque chose de moins."

18. [Macquer], *Dictionnaire de Chymie* (Paris: Lacombe, 1766), 2:170–74.

19. Lavoisier, "Nouvelle theorie," in Fric, "Contribution," p. 160.

20. Cahier lab. 1, pp. 58–59. "j'ay pris une once de cette dissolution que j'ay precipité par mon alkali volatil caustique" "a un peu detonné a la maniere de la poudre mais sans aucun fulmination." "J'ay laissé cet or bien secher sur le filtre et j'ay ete bien etonné de voir ensuitte quil fulminoit aussy bien qu'aucun autre or fulminant."

21. Lavoisier, *Opuscules*, pp. 59–69.

22. Guerlac, *Crucial Year*, pp. 16–17.

23. Gough, "Lavoisier's Early Career," p. 55.

24. Meldrum, *Eighteenth Century Revolution*, p. 27.

25. Cahier lab. 1, p. 60. "Pour decider tout d'un coup et de manier a ne laisser aucune equivoque si la chaux donnoit quelque chose a lalkali caustique a la facon de Meyer ou si au contraire la chaux lui enlevoit quelque chose suivant lhipothese angloise."

26. Lavoisier, "De la nature des eaux d'une partie de la Franch-Comté, de l'Alsace, de la Lorraine, de la Champagne, de la Brie, et du Valois," *Oeuvres de Lavoisier* 2 (Paris: Imprimerie Impériale, 1865), p. 145. Louise Palmer is preparing a doctoral dissertation at Yale University on the scientific activities of Lavoisier between 1764 and 1768, which will give a comprehensive account of his aspirations to become a broad-ranging natural historian during his early career.

27. Lavoisier, "Nature des Eaux," pp. 145–62. For an insightful discussion of the context within which Lavoisier devised his areometer, see Bernadette Bensaude-Vincent, "The Chemist's Balance for Fluids: The Hydrometer and Its Multiple Identities," paper presented at the Conference on "The Role of Experiments and Instruments in the History of Chemistry," Dibner Institute for the History of Science and Technology, Cambridge, Massachusetts, April 11–12, 1996.

28. Cahier lab. 1, pp. 60–61. "diminution de pesanteur specifique occasionnée par la chaux dans une lessive alkaline."

29. Lavoisier, *Opuscules*, pp. 190–93.

30. Cahier lab. 1, p. 62. "cependant cette effervescence quoique fort tumultueux est fort differente de celle qui a lieu avec la terre calcaire ordinaire."

31. Ibid., p. 63. "quelques circonstances ont derangées lappareil le melange s'est fait plus vite que je ne voulois je n'ay pas pu verser les dernieres portions." "Cependant j'ay eu environ 5 pouces 9 lignes dair qui pendant la nuit se sont reduite a un peu moins de 5 pouces."

32. Ibid. "si loperation eut reussy j'aurois eu 6 pouces 1/2 dair cest a dire environ 180 pouces dair cest a dire 1 gros 18 grains."

33. Ibid.

> en quadruplant cette quantite on aura 5 gros pour la pesanteur de lair degagé dans loperation de 18 may la perte du poids a ete de 8 gros ce qui prouve quil y a eu une quantite notable de fluide evaporé ou que lair fixe est plus pesant que lair ordinaire.

34. Ibid. "dou il suit que levaporation du fluide malgré la grande chaleur ne ete que de 20 grains environ."
35. Black, *Magnesia alba*, p. 27.
36. Lavoisier, *Opuscules*, p. 70.
37. Cahier lab. 1. p. 64. "j'ay precipité par un alkali caustique fixe qui ne faisoit avec les acides qu'une effervescence presqu' insensible."
38. Cahier lab. 1, pp. 65–67.
39. Lavoisier, *Opuscules*, p. 262.
40. Cahier lab. 1, pp. 67–68. "jay fait la combinaison peu a peu."
41. Ibid., p. 68.

> en supposant quelle entre pour 2 gros dans la diminution du poids il resteroit 1 gros 1/2 pour le poids de lair degagé pendant la present operation or 1 gros 1/2 dair en ete repondent a un peu moins de 200 pouces cubiques dou il suit que lair fixe est a peupres equiponderable a lair commun que nous respirons.

42. William B. Willcox, ed., *The Papers of Benjamin Franklin*, vol. 20 (New Haven, CT: Yale University Press, 1976), p. 242.

Chapter Eight
The Long Summer Campaign

1. Cahier lab. 1, p. 69.
2. Lavoisier, *Opuscules*, p. i.
3. Ibid., p. 266.
4. Cahier lab. 1, p. 70.

> le degagement dair a eu lieu comme a lordinaire mais ayant donné sur la fin un coupe de feu un peu plus fort qu'a lordinaire le degagement a ete si abondant que j'ay cru la soudure de la cornue de fer fondu . . . j'ay vu que je metois effrayé mal a propos . . . je n'ay pas ete jusqu'au bout. . . . j'ay de lincertitude sur la quantite dair degagé mais j'au lieu de croire quelle a ete de 15 a 16 pouces de mesurées sur la hauteur de ma cloche ce qui donne de 415 a 440 pouces et au moins 1/2 en poids en le supposant equiponderable a lair commun.

5. Ibid.

il y a eu par consequent 2 gros 14 grains de perte

il s'est trouvé dans la boule de fer blanc qui sort du recipient quelques gouttes deau qui netoient ni acides ni alkalines. il en est sorti aussy un peu par le tuyeau montant. elle etoit en vapeurs et se condensoit sur les parois du recipient.

6. Lavoisier, *Opuscules*, pp. 269–70.

7. Cahier lab.1, p. 71. "curieux dexaminer cet air."

cet air se meloit un peu plus promptement que lair fixe davec lair de latmosphere.

. . . y est resté quelques secondes sans paroitre souffrir . . . il ne paroissoit avoir la respiration difficile ni commencement de convulsion S'il eut resté pareil tems dans lair fixe il y fut mort sur le champ.

8. Ibid., p. 72.

persuadé que la combustion du phosphor absorbe lair fixe contenu dans lair ou plutot le soupconnant j'ay pensé que rendant de lair fixe a cet air on pourroit peut etre le rendre air commun . . . j'ay introduit trop dair fixe . . . Ce melange eteignoit egalement une petite bougie meme plus promptement a ce quil ma paru que lair seul dans lequel brule le phosphor.

9. Ibid., pp. 72–74.

il auroit ete bien interessant de rassembler cet acide dans cet etat pour le peser et avoir laugmentation de poids pour vois si elle etoit proportionelle a labsorbtion mais comme cet acide est tres deliquescent il etoit a craindre quil nattira lhumidite de lair pendant meme quon le transvaseroit. pour eviter une partie de cet inconvenient j'ay pensé quil convenoit de faire chauffer le mercure et la cloche affin de faire la transvasion toute chaude.

10. Cahier lab. 1, p. 74; Black, *Magnesia alba*, p. 42; [Macquer], *Dictionnaire*, 2:580–82.

11. The papers which Lavoisier delivered on this, and two further meetings in July 1773, have not survived in their original form, but historians have assumed that their content is the same as part 2 of his *Opuscules physiques et chymiques*, published at the beginning of 1774. See Daumas, *Lavoisier*, p. 30, and Guerlac and Perrin, "Chronology," p. 17. Poirier maintains, however, that Lavoisier wrote the *Opuscules* between September and November of 1773. See Jean-Pierre Poirier, *Antoine Laurent de Lavoisier, 1743–1794* (Paris: Pygmalion, 1993), p. 74. The contents of part 2 can be fitted nicely with the chronology of the laboratory notebooks on the assumption that he read successive installments at these three meetings in the same order that they appear in the book. I have, therefore, made that assumption in the following account.

12. Black, *Magnesia alba*, p. 25.

13. Cahier lab. 1, pp. 75–77.

14. Ibid., pp. 77–78. "oter a leau le gout deau de chaux il falloit encore plus dair fixe."

15. Ibid., p. 79.

> la quantite de 4 onces 1 gros 34 1/2 de soude netoit pas suffisante pour la saturation elle devoit etre de 4 onces 3 gros environ. mais peu importe.

16. Ibid., pp. 79–80. "Comme le barometre est a 28 pouces 1 ligne 1/2 et que le thermometre est a 15 1/2 on ne peut pas se tromper en estimant le pouce cube a 0.46" "lexcedent est de [1 gros, 23 1/2 grains] ‹2 gros 2.42 grains› ce qui est de peu dobjet."

17. Ibid., p. 80. "passer lair de la soude dans la chaux."

18. Lavoisier, *Opuscules*, pp. 187–88.

19. Ibid., pp. 189–201.

20. Ibid., pp. 201–2.

21. Ibid., pp. 203–13.

22. Ibid., p. 213.

23. Cahier lab. 1, pp. 80–81. "de lair attiré pendant la nuit par lalkali caustique."

24. Ibid., pp. 81–82.

25. Ibid., p. 82.

26. Lavoisier, "Nouvelle theorie," Fric, "Contribution," p. 159.

27. [Macquer], *Dictionnaire*, 1:16, 2:46.

28. Lavoisier, *Opuscules*, p. 133.

29. Ibid., pp. 248–49; Cahier lab. 1, p. 83. "je fus tres surpris de voir que la perte de poids netoit plus que de 54 grains" "la dissolution avoit augmenté de poids dun gros."

30. Lavoisier, *Opuscules*, pp. 248–50; Cahier lab. 1, pp. 84–85. "couleur jaune asses belle."

31. Lavoisier, *Opuscules*, pp. 252–53; Cahier lab. 1, p. 85. "un peu moins sec" "empeche de juger de leffet de la dessication."

32. Cahier lab. 1, pp. 88–89. "gonflement et effervescence."

33. Ibid., pp. 89–90.

34. Lavoisier, *Opuscules*, p. 217. I have reconstructed the memoir Lavoisier read on July 17 on the assumption that it is contained in chapters 2–4 of the *Opuscules.* Some sections of these chapters contain sets of experiments performed later, and I have assumed that Lavoisier added these sections afterward without modifying the sections based on the experiments performed during the two weeks prior to his reading. There is some risk that in doing so, I attribute to Lavoisier ideas or transformations of the experiments as recorded in his laboratory record that he actually came to 2–3 months later. Such displacement, however, would not seriously distort the general picture of the evolution of his thought and method.

35. Lavoisier, *Opuscules*, pp. 218–22.
36. Ibid., pp. 220–21.
37. Ibid., pp. 221–22.
38. Ibid., pp. 222–26.
39. Ibid., p. 60.
40. Ibid., pp. 227–30.
41. Ibid., pp. 245–46.
42. Ibid., p. 43.
43. Ibid., pp. 247–48.
44. Ibid., pp. 252–53.
45. Ibid., p. 254.
46. Donovan, *Antoine Lavoisier*, pp. 49–55.
47. Lavoisier, "Nature de l'eau," pp. 151, 152.

Chapter Nine
The End of the Beginning

1. Lavoisier, *Opuscules*, p. 246.
2. Black, *Magnesia alba*, p. 34.
3. Cahier lab. 1, p. 91.
4. Ibid., p. 92. "comme mon pese liqueur dargent netoit pas fait pour les liqueurs plus legeres que l'eau, il tomboit au fond."
5. Ibid., p. 93.
6. Ibid., p. 94.
7. Ibid. "trouvé de 6 onces 3 gros 70 grains ‹au moins› Ce qui fait une augmentation de 6 grains" "doit etre attribué a la combustion de 5 grains de phosphore." "je restraindrai laugmentation a 6 grains en tout."
8. Ibid., pp. 94–96. "je l'ay attribué jusqu'cy a la fixation de la meme air mais ne pourroit-il pas arriver que ce fut la seule humidite de lair que fut absorbé et que lair ainsy privé de son humidite ne diminat de volume" "absorbtion deau ou si une autre substance contenue dans lair y contribuoit." "labsorbtion de lair a ete a peu pres la meme." "la hauteur du mercure ne varioit pas" "environ 2 pouces dair . . . par chaque grain de phosphor meme un peu plus . . . labsence ou le presence de leau n'augment donc ni diminue labsorbtion de lair."
9. Lavoisier, *Opuscules*, p. 240.
10. Ibid., p. 96. "une legere effervescence" "tres privé dair fixe mais le precipité soit a grande eau soit autrement a ete presque nul . . . on peut donc regarder que lalkali volatil tres caustique ne precipite pas la terre calcaire."
11. Ibid., pp. 96–97. "tous les precipites sont moins pesant quil ne devoient letre ce qui me fait presumer que le feu qui a servi a les dessecher etoit trop fort et a volatilisé soit du mercure coulant soit du precipité."
12. Ibid., pp. 97–98. "sans addition" "precipite de mercure par la craye" "dilaté comme a lordinaire" "quelques atomes" "le feu ayant fait fondre ma cor-

nue a la fin de loperation je nay pu connoitre exactement la quantite dair produitte ou absorbée."

13. Ibid., pp. 98–99. "adapté un balon percé dun petit trou" "dans le commencement de loperation il est passé environ 24 ou 30 grains dun phlegm."

14. Ibid., p. 102.

> cependant avec des precautions je suis parvenu a avoir une quantite de plomb fort-considerable fondu le rest etoit partie en grenouilles partie en chaux. je nay pas osé aller plus loin dans la crainte de calciner trop de plomb.
>
> . . . je reduire par estimation a 5 onces, 3 gros, 60 grains.

15. Lavoisier, *Opuscules*, p. 273.

16. Cahier lab. 1, pp. 103–4. "par la combinaison de lair fixe qui s'emanoit du bas par la reduction" "si comme le minium ainsy chargé dair fixe etoit devenu irreductible." "mais il netoit pas aise den bien juger."

17. Ibid., pp. 103v–4v. "Cette experience merite detre recommencée." "Nouvelle reduction de minium ou chaux de Plomb."

18. Ibid., pp. 103, 106. "j'ay poussé le feu a peu pres jusquau point ou il ne passoit plus dair. le degagement a ete plus considerable qu'a l'ordinaire" "je crois avoir fait partir tout lair fixe ce qui ne metoit pas arrivé les fois precedentes." Lavoisier apparently began to record the experiment on p. 103, broke off, and began again on p. 106. It is possible that he began the experiment itself, stopped it, and began again, but if so, he did not say so. Lavoisier, *Opuscules*, p. 268.

19. Ibid., pp. 48v–49, 99–101.

20. Ibid., p. 106.

21. Cahier lab. 1, pp. 106–7.

22. Ibid., p. 107. "il y a eu toujours un peu dair perdu dans chacune des operations de sorte que je supposerai le contenance du bocal de 66 pouces. on voit donc que 126 pouces en passant par leau de chaux ont ete reduits dans cette operation a 66 ce qui fait une reduction a moitié."

23. Ibid., p. 106v. "une petite erreur dans loperation cy contre parce quil s'est echappe de l'air de sorte quil na reelement pas ete employé plus de 121 pouces."

24. Ibid., pp. 107–8. "eteignoit encore la chandelle comme auroit fait de lair fixe" "en moins dune minute il est tombé sur le coté. je l'ay retiré mais il netoit plus tems il etoit mort."

25. Ibid., p. 108. "A peri en un tiers de minute."

26. Ibid., "beaucoup de molecules."

27. Lavoisier, *Opuscules*, pp. 268–69.

28. Ibid., p. 271.

29. One small point in favor of this view is that the weight of the residue of the reduction of minium that Lavoisier reported was that based on the first weighing he made, not the second weighing made "after several days." See Lavoisier, *Opuscules*, p. 268.

30. Ibid., p. 254.
31. Lavoisier, "Pli cachete," in Fric, *Correspondance*, 2:389.
32. Lavoisier, *Opuscules*, pp. 254–55.
33. Ibid., pp. 256–70.
34. Ibid., pp. 269–71.
35. Ibid., pp. 272–73.
36. Ibid., pp. 279–80.
37. Ibid., pp. 280–81.
38. Ibid., p. 254.
39. Ibid., p. i.
40. Trudaine de Montigny to Lavoisier, August 2, 1773, Fonds Lavoisier, Chabrol fol 2, dos. 3, Archives de l'Académie des sciences, Paris.

> je suis enchanté de voir que vous compter publier sur l'air fixe. je ne puis qu'etre trés flatté que vous voulez bien parler de moi. mais je vous supplier instament de ne pas me donner des éloges sur lesquelles votre amitie pour moi vous égare surement.

41. Ibid.

> je vous prie instament, Monsieur, de vouloir bien faire une expérience qui me paroitre jetter un grand jour sur la theorie qu'on peut tirer de toutes celles qui ont été ecrites sur les differentes especes d'air. vous nous avez prouvé par des expériences trés certaines que les metaux par leur calcination absorboient une grande quantité d'air. je voudrois qu'il fut possible d'abord d'éprouver si sous le recipient de la machine pneumatique ils soient susceptible de calcination. 2^0 Si on les faisoit calciner dans une quantité d'air telle qu'elle ne fut pas suffisante pour saturer . . . le metal a calciner si la totalité de l'air seroit absorber. 4^0 enfin il devoit trés essentiel de pouvoir se procurer cet air residu de celui absorbé par la calcination des metaux et d'essuyer s'il peut servir a la respiration des animaux et si la bougie ne s'y eteint pas. Dans le dernier cas que je soupconne de voir avoir lieu il devoit bien essential de mesurer par un moyen quelconque la longueur de la flamme de cette bougie dans l'air commun et dans celui et le tems que cette flamme durreroit. je crois que toutes ces expériences jetteroient un grand jour sur la theorie de l'air fixe.

42. Poirier, *Lavoisier*, p. 74.
43. In this and the following paragraphs, my reconstruction of the treatise as Lavoisier composed it in August is based on plausible, but not conclusive, assumptions. Where the structure of a chapter appears to be based largely on experiments found in the laboratory record prior to August 7, but includes some additional experiments, I have assumed that later experiments were inserted with only minor modifications to the remaining text of the chapter.
44. Lavoisier, *Opuscules*, pp. 333–36; Cahier lab. 1, p. 113. "cest un peu plus

que le poids de lair absorbé ce qui prouve quil y a eu absorbtion deau qui y a contribué."

45. Ibid., p. 114. "dun point apre quoi le phosphor sest eteint."

46. Ibid. "il est clair que dans cette experience il n y a pas eu plus de phosphor brulé que dans le precedente meme quelque chose de moins."

47. Lavoisier, *Opuscules*, pp. 327–32.

48. Ibid., p. 333.

49. Ibid., pp. 335–36.

50. Ibid., pp. 337–41.

51. Daumas, *Lavoisier*, p. 31.

Chapter Ten
Mopping-Up Operations

1. Cahier lab. 2, pp. 9–16. "Verification des experiences de M. Lavoisier sur la fixation de l'air dans les corps et sur le fluide elastique qui s'en degage dans plusieurs circonstances."

2. Ibid., pp. 17–19.

3. Ibid., pp. 20–22. "*immediatement* après qu'on a en mis du feu dans le fourneau, leau a baissé a peu près proportionellement au volume de lair contenue dans la cornue ensuite le fond de la cornue commençant a rougir il y a eu un abaissement considerable de leau en assez prompt pour etre sensible a la vue."

4. Ibid., pp. 21–22.

5. Ibid. "n'eteignoit . . . a beaucoup pres aussi promtement" "devenoit plus efficace pour eteindre la flamme" "comme mort en un moment."

6. Lavoisier, *Opuscules*, p. 1.

7. Cahier lab. 2, p. 23. "plus lentement que celui de la reduction du minium."

8. Ibid.

> air de la respiration—introduit par un tube de verre dans de leau de chaux la precipité tres promtement
>
> de lair de l'atmosphere introduit dans de leau de chaux par un soufflet ne la pas troublée ni precipité de meme.

9. For the further development of this interest, see Frederic L. Holmes, *Lavoisier and the Chemistry of Life* (Madison: University of Wisconsin Press, 1985).

10. Cahier lab. 2, p. 24.

11. See Frederic L. Holmes, "The Boundaries of Lavoisier's Chemical Revolution," *Revue d'Histoire des Sciences* 48 (1995): 9–48.

12. Macquer to Bergman, January 17, 1774, *Torbern Bergman's Correspondence*, ed. Göte Carlid and Johan Nordström, vol. 1 (Stockholm: Almqvist and Wiksell, 1965), p. 244.

13. Cahier lab. 1, pp. 116–20.

14. Cahier lab. 2, pp. 4–5.

15. Ibid., pp. 25–30.

16. Ibid., p. 31.

17. Ibid., p. 32.

18. Ibid. "Ce qui revient tres exactement aux resulte que jay donné dans mon memoire."

19. Ibid., p. 43. "n y a pas ete precipité. Cependant apres une asses longue agitation elle a louchi un peu mais cette circonstance ne venoit elle de lair fixe produit par la bougie qui avoit ete introduitte."

20. Ibid., p. 44.

21. Ibid., p. 45. "1 grain [et demi au moins] ‹trois quart au moins›."

22. Ibid.

23. Ibid., pp. 46–48.

24. Lavoisier, *Opuscules*, p. 293.

25. Ibid., p. 290.

26. Ibid., p. 293.

27. Ibid., p. 346.

28. Cahier lab. 2, p. 57. "M. de trudaine m'ayant fait naitre quelquinquietude sur le degagement de lalkali volatil du sel ammoniac par les terres calcaires precipités sous forme caustique et non caustique j'ay repeté de nouveau toutes les experiences."

29. Lavoisier, *Opuscules*, pp. 347–49.

30. Ibid., pp. 350–52.

31. Meldrum, *Revolution in Science*, p. 17; Poirier, *Lavoisier*, p. 75.

32. De Trudaine, Macquer, Le Roy, and Cadet, "Rapport fait a l'Académie," printed in Lavoisier, *Opuscules*, pp. 365–66.

33. Ibid., pp. 368–69.

34. Ibid., p. 371.

35. Ibid., p. 372.

36. Ibid., p. 373.

37. Ibid., pp. 375–76.

38. Ibid., p. 376.

39. Lavoisier, *Opuscules*, title page; Daumas, *Lavoisier*, p. 32.

40. Lavoisier, *Opuscules*, pp. iii–viii.

41. Poirier, *Lavoisier*, p. 75.

42. Fric, *Correspondance*, 2:398–401.

43. Cahier lab. 1, p. 3. "reprendre tout ce travail qui m'a paru fait pour occasioner une revolution en Phisique et en chimie."

44. Ibid., p. 411.

45. Ibid., pp. 411–12.

46. Reproduced in Bernadette Bensaude-Vincent, "A View of the Chemical Revolution through Contemporary Textbooks: Lavoisier, Fourcroy and Chaptal," *British Journal for the History of Science* 23 (1990): 456–57.

47. Donovan, *Antoine Lavoisier*, pp. 47–49.

CHAPTER ELEVEN
CONCLUSION

1. Meldrum, *Revolution in Science*, p. 32.

2. Maurice Daumas, *Lavoisier* (Paris: Gallimard, 1941), pp. 99–100.

3. A. R. Hall, *The Scientific Revolution, 1500–1800* (Boston: Beacon Press, 1954), p. 332.

4. Meldrum, *Revolution in Science*, p. 34.

5. Berthelot, *Révolution chimique*, p. 3.

6. Howard Gardner, *Creating Minds* (New York: Basic Books, 1993), pp. 43–44.

7. Trudaine to Lavoiser, August 2, 1773, Fonds Lavoisier, AAdS. "qui cultivent les sciences et qui les avancent."

CHAPTER TWELVE
BEFORE AND AFTER 1773: LAVOISIER STUDIES

1. Robert E. Kohler, "Gerald L. Geison, *The Private Science of Louis Pasteur*," *Isis* 87 (1996): 334.

2. Roger Hahn, "Lavoisier et ses collaborateurs: une équipe au travail," in *Il y a 200 Ans Lavoisier: Actes du Colloque organisé à l'occasion de la mort d'Antoine Lavoisier*, ed. C. Demeulenaere-Douyère (Paris: Tec & Doc Lavoisier, 1995), pp. 55–63.

3. Bensaude-Vincent, *Lavoisier*, p. 28.

4. Edouard Grimaux, *Lavoisier: 1743–1794* (Paris: Félix Alcan, 1888), pp. 11–34.

5. J. B. Gough, "The Origins of Lavoisier's Theory of the Gaseous State," in *The Analytic Spirit: Essays in the History of Science in Honor of Henry Guerlac*, ed. Harry Woolf (Ithaca: Cornell University Press, 1981), pp. 15–39.

6. Marco Beretta, *The Enlightenment of Matter: The Definition of Chemistry from Agricola to Lavoisier* (Canton, MA: Science History Publications, 1993), pp. 159–69.

7. Donovan, *Antoine Lavoisier*, p. 6.

8. Evan M. Melhado, "Chemistry, Physics, and the Chemical Revolution," *Isis* 76 (1985): 195–211.

9. Beretta, *Enlightenment*, pp. 163–66.

10. Bensaude-Vincent, *Lavoisier*, pp. 68–79.

11. See, for example, Guerlac, *Antoine-Laurent Lavoisier*, p. 52; Poirier, *Lavoisier*, p. 8.

12. The Ph.D. dissertation being prepared at Yale University by Louise Palmer will describe in depth Lavoisier's work in mineralogy, geology, and chemistry during the period 1764–67.

13. Charles Coulston Gillispie, *The Edge of Objectivity: An Essay in the History of Scientific Ideas* (Princeton: Princeton University Press, 1960), p. 321.

14. Jan Golinski, *Science as Public Culture: Chemistry and Enlightenment in Britain, 1760–1820* (Cambridge: Cambridge University Press, 1992), pp. 129–52.

15. Ibid., pp. 137, 142n, 145.

16. Steven Shapin and Simon Schaffer, *Leviathan and the Air Pump: Hobbes, Boyle, and the Experimental Life* (Princeton: Princeton University Press, 1985), esp. pp. 40–49.

17. For a more extended comment on the same subject, see Frederic L. Holmes, "Do We Understand Historically How Experimental Knowledge Is Acquired?" *History of Science* 30 (1992): 119–36.

18. Hans-Jörg Rheinberger, "Experiment, Difference, and Writing: I. Tracing Protein Synthesis," *Stud. Hist. Phil. Sci.* 23 (1992): 309; Rheinberger, "Experimentation and Orientation: Early Systems of In Vitro Protein Synthesis," *Journal of the History of Biology* 26 (1993): 465.

19. Bensaude-Vincent, *Lavoisier*, pp. 129–38, 349–58.

20. Aldo Mielo, ed., "Una lettera di A. Lavoisier a J. Black," *Archeion* 25 (1943): 238–39.

* *Index* *